Research Advances in Syngas

Edited by Abrar Inayat and Lisandra Rocha-Meneses

Published in London, United Kingdom

Research Advances in Syngas
http://dx.doi.org/10.5772/intechopen.1000350
Edited by Abrar Inayat and Lisandra Rocha-Meneses

Contributors
Abrar Inayat, Dixon Domfeh, Farrukh Jamil, Hardik Y. Desai, Iram Liaqat, Jean-Louis Kindler, Kalsoom Tahir, Lisandra Rocha-Meneses, Martha Siles Camacho, Mira Al Balushi, Nasiru Zubairu, Nauman Aftab, Saba Mahboob, Sikander Ali, Tamer M. Abdellatief, Valeria Sandoval Cossío

First published in London, United Kingdom, 2024 by IntechOpen
IntechOpen is the global imprint of INTECHOPEN LIMITED, registered in England and Wales, registration number: 11086078, 5 Princes Gate Court, London, SW7 2QJ, United Kingdom

British Library Cataloguing-in-Publication Data
A catalogue record for this book is available from the British Library

Additional hard and PDF copies can be obtained from orders@intechopen.com

Research Advances in Syngas
Edited by Abrar Inayat and Lisandra Rocha-Meneses
p. cm.
Print ISBN 978-1-83769-682-6
Online ISBN 978-1-83769-681-9
eBook (PDF) ISBN 978-1-83769-683-3

For EU product safety concerns:
IN TECH d.o.o., Prolaz Marije Krucifikse Kozulić 3, 51000 Rijeka, Croatia,
info@intechopen.com or visit our website at intechopen.com.

Meet the editors

Dr. Abrar Inayat, a talented researcher with more than 13 years of experience in sustainable and renewable energy and chemical engineering, holds a Ph.D. in Chemical Engineering with a focus on Renewable Energy from Universiti Teknologi PETRONAS, Malaysia. He earned his MEng from Mälardalens University, Sweden, specializing in Sustainable Energy Systems. His undergraduate studies include a BSc in Chemical Engineering. Dr. Inayat's prolific research is highlighted by the publication of more than 200 research papers. Recognized as one of the world's top 2% of scientists in the energy category by Stanford University, USA, he is also the recipient of the Best Researcher Award at Universiti Teknologi PETRONAS. His outstanding contributions have been recognized with various international awards.

Dr. Lisandra Rocha-Meneses received her Ph.D. in Engineering Sciences from the Estonian University of Life Sciences. She holds a BSc in Renewable Energy and an MSc in Management and Nature Conservation, both from the University of Azores, Portugal. Dr. Lisandra was awarded in 2024 among 60 Impactful Women in the Middle East in Climate and Sustainability. She is the recipient of the award L'Oréal Baltic program "Women in Science": 2020 Young Talent for Estonia. She has 8 years of international experience as a researcher in countries such as Canada, Sweden, Australia, Estonia, the United Arab Emirates, and Ireland. Dr. Meneses has also attended many international courses, training, and conferences. She is an associate editor of Agronomy Research and has been involved in the organization of scientific conferences and workshops since 2013.

Contents

Preface

For centuries, fossil fuels have been the driving force behind civilization, leaving an impact not only on our progress but also on our planet. However, as we confront the looming threat of climate change, we find ourselves at a juncture. We must decide whether to continue down the path of carbon energy or forge a new way forward that is illuminated by cleaner and more sustainable energy sources.

This book, *Research Advances in Syngas*, delves deeply into this crossroads. It explores syngas—a mixture produced from various sources like coal, biomass, and even waste—and presents an intriguing glimpse into a future where energy can be both powerful and environmentally responsible.

Within the pages of this book you will embark on a journey through the frontiers of syngas research. The opening chapter sets the stage by exploring the evolution of syngas production—from being a relic to potentially becoming a champion for clean energy.

Chapter 2 then delves into the core issue: how we can replace environmentally harmful energy sources with syngas derived from accessible and sustainable feedstocks. This exploration is crucial, as it paves the way for an energy economy that relies less on fuels and is more in harmony with our environment. Understanding the relationship between feedstock and syngas is crucial for optimizing production and minimizing the impact on the environment.

Chapter 3 explores the world of syngas production by focusing on developing catalysts that are also environmentally friendly.

Chapter 4 closely examines the factors that influence natural gas prices in the United States. This insightful analysis reveals how market dynamics and energy choices interact, providing context for navigating the landscape of syngas adoption.

Chapter 5 delves into a toolbox that explores how applying supply chain strategies can improve the performance of syngas networks. By optimizing logistics and streamlining processes we can ensure cost-effective delivery of this energy source.

Finally, Chapter 6 ventures into the cutting-edge field of carbon-hydrogen production through steam induced vapolysis. This revolutionary technology not only generates energy but also actively removes carbon dioxide from the atmosphere, offering a glimpse of hope for a truly sustainable future.

Research Advances in Syngas is more than a compilation of discoveries; it is a road-map towards a cleaner and more prosperous future. Each chapter delves into the complexities of syngas production and explores its economic, environmental, and societal implications. Whether you have experience in research, care deeply about the environment, or aspire to promote energy, this book welcomes you to participate in the discussion. Let us explore the world of syngas innovation together and shed light on the journey towards a more sustainable future.

Dr. Abrar Inayat
University of Sharjah,
Sharjah, United Arab Emirates

Dr. Lisandra Rocha-Meneses
Wrocław University of Environmental and Life Sciences,
Department of Applied Bioeconomy,
Waste and Biomass Valorization Group (WBVG),
Wrocław, Poland

Chapter 1

Introductory Chapter: Synthesis Gas Production – History and Current Scenario

Abrar Inayat, Farrukh Jamil, Lisandra Rocha-Meneses and Tamer M.M. Abdellatief

1. Introduction

In the ever-changing landscape of energy research, syngas, also known as synthesis gas, stands out as a crucial linchpin in the tireless quest for sustainable and efficient energy solutions. This inaugural chapter not only serves as the gateway to an extensive exploration of the latest advancements in syngas research but also marks the inception of a profound investigation into its diverse applications, the challenges it presents, and the promising horizons it opens for the creation of a sustainable energy future.

2. The genesis of syngas

The term "syngas" encompasses a gaseous mixture primarily consisting of carbon monoxide (CO) and hydrogen (H_2). Originating in the early nineteenth century, a period characterized by the burgeoning industrial revolution and an unquenchable thirst for innovative energy solutions, the genesis of syngas can be traced back to pioneering experiments in gasification processes [1]. These researchers sought to comprehend the transformative potential of converting carbonaceous materials into a gaseous mixture.

Sir William Grove's foundational work in the 1830s laid the bedrock for the discovery of the gasification process, a precursor to syngas production. Grove's experiments in electrolysis and subsequent synthesis of hydrogen and carbon monoxide provided the initial insights into manipulating gases for energy applications. Progressing through the century, scientists such as Franz Fischer and Hans Tropsch made seminal contributions, refining gasification techniques and exploring catalytic processes integral to syngas production.

The end of the nineteenth and beginning of the twentieth centuries saw the inception of coal gasification plants, representing a notable achievement in the industrial utilization of syngas. These urban facilities were dedicated to producing syngas for lighting and heating purposes. The Fischer-Tropsch synthesis process in the 1920s further enhanced the significance of syngas, enabling the synthesis of liquid hydrocarbons from carbon monoxide and hydrogen. World War II heightened the strategic importance of syngas as a synthetic fuel source during periods of petroleum scarcity. Substantial investments during this time led to advancements in gasification technology and the exploration of various feedstocks, including coal and biomass [2].

The latter part of the twentieth century witnessed a renewed focus on syngas, spurred by the exploration of cleaner and more sustainable energy alternatives. Researchers delved into innovative production methods, particularly exploring biomass gasification and natural gas reforming. The introduction of advanced analytical tools and computational modeling broadened our understanding of syngas chemistry and catalysis. In the twenty-first century, the genesis of syngas expanded beyond traditional fossil fuels to encompass a wider range of feedstocks, including biomass, municipal waste, and even carbon dioxide. The ongoing pursuit of cleaner energy solutions has reinvigorated syngas as a crucial link between conventional and renewable energy sources, positioning it at the forefront of state-of-the-art research in the global pursuit of sustainable energy.

The evolution of syngas unfolds as a testament to human innovation and adaptability, charting its course from the industrial revolution to the cutting edge of modern energy research. As we immerse ourselves in the complex tapestry of syngas development, we unveil a compelling history that showcases the transformative metamorphosis of an embryonic idea into a central force shaping the contemporary energy background. This narrative stands as a testament to the power of innovative thinking and the indomitable spirit that propels breakthroughs in the dynamic landscape of energy exploration.

3. Significance and versatility

The significance and versatility of syngas extend far beyond its chemical composition, positioning it as a transformative and adaptable resource in contemporary energy research. Derived from diverse feedstocks like biomass, natural gas, and coal, syngas emerges as a flexible building block with a multitude of applications, spanning power generation, chemical synthesis, and various industrial processes. Its unique capacity to be tailored for specific needs establishes syngas as a crucial bridge connecting traditional fossil fuels with renewable energy sources, offering a flexible and sustainable alternative.

In essence, syngas is not merely a byproduct of gasification processes; rather, it stands as a cornerstone for synthesizing a diverse array of valuable fuels and chemicals. As the energy landscape evolves toward cleaner and more sustainable solutions, the significance of syngas continues to grow, solidifying its role as a linchpin in the pursuit of a resilient and environmentally conscious energy future [3].

4. Milestones and breakthroughs

Charting the trajectory of syngas research reveals a series of milestones and breakthroughs that have shaped its contemporary status. From catalytic advancements to reactor technologies, researchers have navigated uncharted territories to push the boundaries of knowledge. This section highlights key breakthroughs, showcasing the ingenuity that has propelled syngas into the forefront of scientific inquiry. By acknowledging past achievements, we gain insights into the foundations upon which current research is built.

Milestones and breakthroughs in syngas production have charted a transformative course, marking key epochs in the evolution of this versatile energy source. The seminal work of early twentieth century scientists Franz Fischer and Hans Tropsch established the foundation with the development of the Fischer-Tropsch synthesis process, enabling the synthesis of liquid hydrocarbons from syngas. Subsequent breakthroughs in the mid-century saw the strategic application of syngas during World War II, emphasizing its importance as a synthetic fuel source in times of petroleum scarcity. Advancements in gasification technologies, particularly with the rise of coal gasification plants, contributed to syngas becoming a ubiquitous fuel for lighting and heating in urban centers. In recent decades, breakthroughs in catalyst development, reactor design, and computational modeling have elevated syngas production to new heights, enabling researchers to optimize efficiency, reduce environmental impact, and explore innovative applications across various sectors. These milestones collectively underscore the dynamic and enduring journey of syngas, from early breakthroughs to contemporary innovations, shaping its profound impact on the landscape of energy production and utilization.

5. Challenges in syngas research

While the potential benefits of syngas are extensive, researchers are confronted with equally massive challenges. This multifaceted landscape involves a complex interplay of technical, economic, and environmental considerations, with feedstock variability, gasification efficiency, and the pursuit of optimal catalysts standing out as significant hurdles. This section meticulously dissects these challenges, offering a nuanced exploration of the factors shaping the current state of syngas research while also delving into the opportunities they present for innovation and the breakthroughs required to surmount them.

The terrain of syngas research is marked by challenges that demand both ingenuity and resilience from researchers, encompassing technical intricacies, economic considerations, and environmental concerns that collectively define the trajectory of contemporary investigations into syngas applications. A central hurdle emerges in dealing with the diverse and variable feedstocks derived from biomass, natural gas, or coal. These factors introduce a multifaceted set of variables that influence both the efficiency of gasification and the quality of the produced syngas. Overcoming the challenge of ensuring uniformity and dependability across a range of diverse feedstocks proves to be a significant obstacle, demanding the implementation of advanced technologies and adaptive strategies. The efficiency of gasification poses a crucial challenge in syngas research, with the ongoing objective of optimizing processes to convert feedstocks into syngas while minimizing energy loss.

Researchers engage with the complexities of reactor design, temperature control, and catalyst development, striving to attain elevated efficiency levels and diminish environmental footprints. The search for ideal catalysts stands out as another substantial challenge in syngas research. Catalysts hold a crucial role in improving reaction rates and selectivity during syngas production; however, the intricate puzzle lies in identifying catalysts that are both economically viable and environmentally sustainable [4]. To address this complexity, researchers delve into the exploration of novel materials and formulations, aiming to strike a delicate balance between efficiency, cost-effectiveness, and minimizing environmental impact.

Considerations related to economics carry significant weight, necessitating the development of scalable and economically viable methods for syngas production. The challenges of managing both capital and operational costs, along with the integration of syngas into existing industrial processes, call for strategic solutions to ensure economic feasibility [5].

Simultaneously, environmental sustainability is of paramount concern, given the cleaner alternative that syngas presents. Despite this, the environmental impact of production processes demands meticulous attention. In response, researchers are actively engaged in the development of environmentally friendly gasification methods and the exploration of carbon capture and utilization techniques.

In the dynamic field of syngas research, these challenges act as catalysts for fostering innovation and exploration. Each obstacle becomes an opportunity for researchers to expand the frontiers of knowledge, devising solutions that not only address immediate challenges but also contribute to the development of a sustainable and resilient energy future. The collective efforts of the scientific community in navigating these challenges play a pivotal role in shaping the transformative potential of syngas within the broader landscape of global energy sustainability.

Author details

Abrar Inayat[1,2]*, Farrukh Jamil[2,3], Lisandra Rocha-Meneses[4]
and Tamer M.M. Abdellatief[5,6]

1 Department of Sustainable and Renewable Energy Engineering, University of Sharjah, Sharjah, United Arab Emirates

2 Biomass and Bioenergy Research Group, Center of Sustainable Energy and Power Systems, Research Institute of Science and Engineering, University of Sharjah, Sharjah, United Arab Emirates

3 Department of Chemical Engineering, COMSATS University Islamabad (CUI), Lahore, Pakistan

4 Wrocław University of Environmental and Life Sciences, Department of Applied Bioeconomy, Waste and Biomass Valorization Group (WBVG), Wrocław, Poland

5 Sustainable Energy and Power Systems Research Centre, RISE, University of Sharjah, Sharjah, United Arab Emirates

6 Faculty of Engineering, Department of Chemical Engineering, Minia University, EL-Minia, Egypt

*Address all correspondence to: ainayat@sharjah.ac.ae

References

[1] Abdoulmoumine N et al. A review on biomass gasification syngas cleanup. Applied Energy. 2015;**155**:294-307

[2] Hagos FY, Aziz ARA, Sulaiman SA. Trends of syngas as a fuel in internal combustion engines. Advances in Mechanical Engineering. 2014;**6**:401587

[3] Rauch R, Hrbek J, Hofbauer H. Biomass gasification for synthesis gas production and applications of the syngas. WIREs Energy Environment. Advances in Bioenergy. 2014;**3**:343-362

[4] Ren J et al. Recent advances in syngas production from biomass catalytic gasification: A critical review on reactors, catalysts, catalytic mechanisms and mathematical models. Renewable and Sustainable Energy Reviews. 2019;**116**:109426

[5] Wilhelm DJ et al. Syngas production for gas-to-liquids applications: Technologies, issues and outlook. Fuel Processing Technology. 2001;**71**(1):139-148

Chapter 2

Development of Syngas Using Feedstock for the Replacement of Expensive Energy Resources

Saba Mahboob, Kalsoom Tahir, Sikander Ali, Iram Liaqat and Nauman Aftab

Abstract

Due to the increasing demand for petroleum use as fuel, there has been a focus on the production of fuel that has a huge possibility for long-term energy sustainability. Synthetic gas (Syngas) is generated by means of a thermochemical technique known as gasification, which converts carbonaceous feedstocks (biomass, crude oil residuum, municipal waste, petroleum, and coal) to syngas. It contains carbon monoxide and hydrogen as the key elements of inflammable gas. It is widely used for gas lighting in coal gasification method before availability of electric lighting, gas turbine fuel, raw material for liquid fuels and the synthetic natural gas, and anode gas of solid oxide fuel cells. It is synthesized either through the gasification of plant-based biomass or pyrolysis of waste. This chapter will focus on the information, which has been rounded up over the last decades on syngas properties, sources, and production at a competitive advantage, as well as application and future technological advancement.

Keywords: synthetic gas, gasification, pyrolysis, thermochemical process, petroleum product, sustainable energy source

1. Introduction

Syngas is also known as the "synthesis gas" and producer gas that could be synthesized from many kinds of resources, which consist of carbon. Energy resources such as biomass, coal, plastics, municipal waste, and other comparable materials. Synthesis gas consists of a mixture of H_2, CO, and CO_2 that could be utilized as a potential intermediate in the transformation of biomass into fuel. Raw syngas also carries mostly notable amounts of CO_2 and H_2O. It is also a short name used for a gasification product, mainly from waste biomass. All of the chemicals and goods that the petrochemical sector today produces are constructed from syngas. Possibly, decisive and the demanding applications of the synthesis gas as precursor for the liquid fuel.

Biomass is a CO_2 neutral option for energy production, and the potential for the conversion of biomass into energy is expanding quickly [1]. According to Maniatis, energy from biomass based on short rotation forestry (SRF) and other energy crops

IntechOpen

performed valuable contribution toward achieving the objectives of the Kyoto agreement in reducing the greenhouse gas emissions and the problems related to climate change [2]. Furthermore, many biomass technologies are present for conversion of biomass to energy. These technologies can change raw biomass into a variety of gaseous, liquid, or solid materials that can then be used for the generation of energy.

This conversion can be performed in three ways: thermochemical (break down of biomass at high temperature), biochemical (break down of biomass in the presence of microorganism or enzymatic processes), and chemical (oils from biomass can be chemically converted into a liquid fuel) [3]. Over the last few years, there have been many developments in the science and technology of thermochemical biomass conversions. Incineration, gasification, and pyrolysis conversion are among the established and best available thermochemical technologies [2, 3]. These thermal processes provide sustainable energy sources that are efficient, environmentally acceptable, and cost-effective. Gasification is the thermochemical conversion of a solid biomass to a gaseous fuel through heating in the presence of gasification agent (i.e., air, oxygen, steam, hydrogen, CO_2, or mixtures of these gases).

It is the technological technique, which could convert carbonaceous raw materials such as biomass, coal, and waste in fuel gas that also called as syngas. Syngas got its name from its usage as in-between in synthesis of the SNG (synthetic natural gas). Primarily, the syngas is utilized in synthesis of other chemicals and fuels such as methanol and diesel fuel at higher pressures [4]. Generally, syngas might indicate combination of (i) nitrogen/hydrogen (N_2/H_2), which is utilized to produce ammonia, (ii) carbon monoxide and hydrogen for the production of synthetic hydrocarbons in the gas-to-liquids plants along with methanol and ethanol in the petrochemicals [5], or (iii) carbon dioxide and hydrogen (CO_2/H_2) to produce hydrogen [6, 7].

Synthesis gas is the significant component of chemical processing industries, and the hydrogen has been a major part of the syngas. The composition of the syngas has highly dependent on the raw material and manufacturing technique but can be thought of as mixture of 30–50% carbon monoxide (CO), 25–30% hydrogen (H), 5–15% carbon dioxide (CO_2), and 0–5% methane (CH_4). Syngas produces diesel fuel by the Fischer-Tropsch technique. This method turns combination of hydrogen and carbon monoxide into liquid hydrocarbons by series of reactions. These chemical reactions could occur in existence of metal catalysts at the temperatures, ranging from 150°C to 300°C (302°F–572°F).

Syngas is a fundamental intermediate product all over the chemical industry. Each year, almost 6 EJ of syngas is synthesized throughout the world, which is nearly 2% of the main energy consumption of the world. The global syngas market is dominated by the ammonia industry (largely from the fossil fuels e.g., natural gas, coal, oil, and residues) [7]. There are various applications of syngas such as the synthesis of the hydrogen for usage in the refineries, such as processing of the hydrogen, gas-turbine fuel, anode gas of the solid oxide fuel cells, the raw material for synthetic natural gas (SNG), and methanol synthesis and liquid fuels. Frequently, these applications need progressions for cleaning synthesis gas to remove impurities that arise from the coal, which comprise sulfur compounds, mercury, hydrogen halides, alkali metals, and trace elements [8].

2. Numerous feedstocks as energy source

For eras, coal had been used as prime feedstock for the gasification process then due to the current concern regarding its usage as fossil fuels, and subsequent environmental contaminants, there is interchange to other feedstocks except the coal [9]. In the gasification process, carbonous feedstocks, which would be unused and cast-off, for example, waste biomass can be used as energy sources to produce syngas. Several feedstocks, for example, biomass, other carbonaceous wastes, and crude oil residues could be utilized to their maximum potential [10].

The gasification process could take different feedstocks, but the reactor that used in the gasification process must be chosen on the base of behavior and feedstock properties in the process. When various feedstocks (carbonaceous or hydrocarbonaceous) have been employed in gasification than product (such as syngas) has possibly been more suitable as the energy resource as compared to other feedstocks. Syngas production has been more effective energy source than the direct combustion of feedstock because the syngas can be combusted on the high temperature, utilized in fuel cell, use to produced methanol, used as hydrogen source, and converted into range of the synthesis liquid fuels that are fit to use for diesel engines, gasoline engines, or for the production of wax [11] (**Figure 1**).

2.1 Natural gas

Searching for the alternate energy resources to substitute the petroleum fuels, the natural gas is engrossed attention of numerous scientists and huge quantity of the methane comprised in the natural gas is contemplated as the contribution in the production of high-valued products, for example, synthesis gas and the great purified hydrogen. Although synthesis gas could be generated from feedstock diversity, for example, petroleum, coke, coal, natural gas, and biomass, but lowest- cost routes to produce syngas have been based on the natural gas [12].

Natural gas demand as energy source has been increased gradually. Due to lack of hetero atoms and high H:C ratio make the natural gas as key feedstock for the synthesis fuel, which could swap these which have been characteristically petroleum-derived products. Additionally, novel arenas of gas and oil comprise 8–18% of associated CO_2, and in approximately precise field, allied gas comprises of high CO_2 (such as 79%). In this situation, process such as tri-reforming syndicates two problematic gases

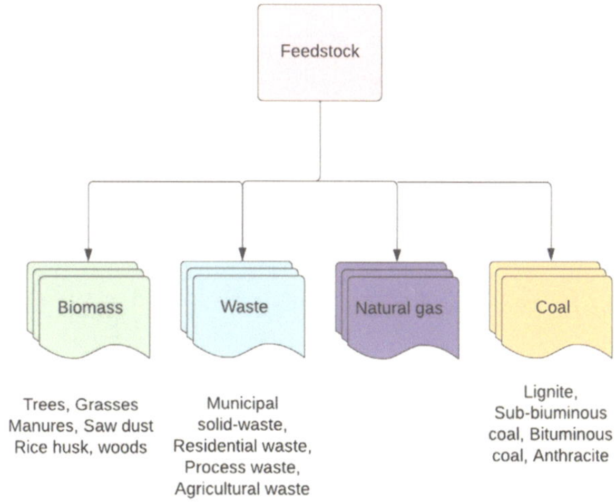

Figure 1.
Different feedstock as energy sources.

of greenhouse (CO_2 or CH_4) to produce synthetic gas for mixture of the cleaned liquid fuel and valued chemicals [13].

Assuming the up-to-date level of the consumption of the natural gas for the world has been upheld, standby should be adequate to last for 64 years. Though, in assessment of the natural gas permanency, influences, for example, upsurge in the annual consumption, finding of novel reservoir, the improvements in the discovery technologies, and the use of the hydrates of natural gas have not been involved. So, the outcome of gas discovery in fitted a slate formation that is balanced over annual consumption due to the discoveries of major fields of natural gas the world's natural gas reserves are generally in an upward trend [14].

2.2 Coal

Chemically, coal is the hydrogen lacking hydrocarbon having an atomic H:C ratio close to 0.8, for example, compared to hydrocarbon derivatives of crude oil, which have an atomic H:C ratio almost equivalent to the 2, and the methane, which has atomic H:C ratio equivalent to the 4. Thus, any process that has been utilized to change the coal into alternate fuel should add hydrogen in natural coal to make coke and ionic product of hydrogen [15].

Various coal types in the organic sedimentary rock, which has been formed from the preservation and accumulation of the plant materials, generally in the marsh environment [7, 8]. It is flammable rock, and laterally with natural gas and oil it has made one of three significant fossil fuels for production of the electrical energy and affords about 40% of the electrical energy production on global basis. It has been studied widely for the conversion into liquid and gaseous along with hydrocarbon feedstocks. Due to his comparative profusion and the constant fuel prices on the market, coal is an important objective for the synthetic adaptation in other forms of fuel, for example, synfuels. It can be gasified along with the additional fuels containing oil, biomass, scrap tires, and the municipal wastes [7, 9, 16–18]. Coal gasification as power-generation technology has been gaining popularity because of global accessibility of coal raw material, in addition to positive environmental issues related to that technology over the other combustion technologies.

Hence, the gasification process provides a versatile and cleanest method to change energy in hydrogen, electrical energy, and other sources of power that contained in coal. Turning of coal in syngas has not been a new idea; actually, the coal gasification method times following prior to World War II. The gasification could process effectively those entire residues and the wastes which have been formed in the processing planted to the improved production of the inflated valued products by the advancement of their crude feedstock [14].

2.3 Biomass

Biomass is the renewable source, which has been established significant consideration due to the conservational contemplations and growing energy demands worldwide [19, 20]. Biomass has been generated through a process called photosynthesis, which includes chemical reactions that occur on the Earth among green plants and sunlight in chemical energy form. For instance,

$$6CO_2 + 12H_2O \rightarrow 6CH_2O + 6O_2 + H_2O \quad (1)$$

There are several forms of the biomass, which could be utilized and replace deprived of irrevocably reducing assets, and usage of the biomass would endure to raise in standing as substitute for sources of fossil fuel and the feedstocks for variety of the products [16, 18, 19]. Up to 35% of the energy needs of emerging nations are met by biomass, which is also equivalent to 13% of the global energy demand. Biomass has been broadly obtainable in the amounts that are adequate to meet world energy demand [21, 22].

Gasification of the biomass resources is the most attractive method to produce a gas that is highly rich in hydrogen. Production of syngas from the biomass has been the most suitable solution for the energy disaster. Depletion of fossil fuel energy in the developing countries can be overcome by utilizing biomass as energy source, for example, ample biomass has been obtainable in the developing countries, and it also renewable [21, 23]. Gasification signifies as effective and the environment friendly technique to produce syngas as biofuel from diverse biomass resources [24] and generate second biofuel generation, for example, hydrogen, ethanol, and methanol [25]. Gasification could be definite as partial biomass combustion, and it can extract till 60–90% of the energy that deposited in biomass [26]. Worldwide, biomass energy is generally integrated in power production system such as the United States commenced incomplete and complete conversion of the traditional power stations to the biomass [27].

2.4 Waste

Focus on the modern management of waste has been considering shift to the production of energy, although handling sustainably of waste. Subsequently, the treatment of waste by gasification has gradually attained waste burning with numerous aids, comprising instantaneous waste management and the production of energy, although dropping landfill volume and dislocating conservative fossil fuels. Just 3% of the entire solid waste has been utilized to produce energy; there is significant capability to utilize the residual waste for the recovery of energy [28].

Gasification of the waste feedstocks is not different to gasification of biomass because they are also hydrocarbons that are used to produce synthesis gas. Waste gasification has been favored over the burning as it provided synthesis gas product, which could be utilized in several applications as contrasting to the hot gases of incineration. Waste gasification provides the even quality synthesis gas from mixed and multifaceted residual waste. It is the solitary option, which could provide multimodal products, for example, heat, cooling, power, liquid, and gaseous fuels in addition to chemicals [29].

3. Production of syngas

3.1 Gasification of biomass

Biomass gasification has one of the attractive approaches to produce gas that are highly rich of hydrogen. It converted carbonous biomass to gases such as hydrogen, carbon monoxide, and carbon dioxide [30]. This method could be attained through a chemical reaction of the feed at 700°C, with the inadequate volume of stream and oxygen. In gasification of biomass, feed has been treated with some portion of combustion to maintain the high gasification reaction temperature. In this process, engendered mixture of the gas has been considered as synthetic gas and producer gas used as fuel [31]. The power produced in gasification of biomass and generating gas combustion could be considered as the renewable resource of energy.

The reactions of biomass gasification have been taken place in gasifier that could be concise as indicated below [23, 32, 33]:

Partially oxidation:

$$C + 1/2\,O_2 \leftrightarrow CO\ \Delta H = -268\ \text{kJ/mol.} \tag{2}$$

Complete oxidation:

$$C + O_2 \leftrightarrow CO_2\ \Delta H = -406\ \text{kJ/mol} \tag{3}$$

Water-gas phase reaction:

$$C + H_2O \leftrightarrow CO + H_2\ \Delta H = 118\ \text{kJ/mol} \tag{4}$$

Boudouard reaction:

$$C + CO_2 \leftrightarrow 2CO\ \Delta H = 170.7\ \text{kJ/mol} \tag{5}$$

Heat requires for the water gas stage and the Boudouard reactions have been performed through partial and complete oxidation reactions, and the complete oxidation reaction provides about 60% of heat necessities in gasification [23].

The biomass gasification has been the probable resource to produce biofuels and energy, especially chemical energy. The gasifier converts renewable energy resources of biomass to the syngas in the gasification process. Produced syngas has been utilized to function of the internal combustion engine. Syngas could be utilized to generate heat energy and electricity through co-generation system. Almost, biomass energy resources are requiring drying first. Later, dry materials have been required for the process of devolatilization [34].

There are following advantages of biomass gasification: (1) Produces a more practical, readily controllable kind of cleaner fuel for the production of thermal energy, as well as electricity, and offers a way to cut back on or dip away with traditional fossil fuels. (2) Gasification enables biomass to be used as a fuel for a variety of power generation technologies, including gas turbines, fuel cells, and reciprocating engines. (3) It is possible to gasify a wide range of biomass materials, many of which would be challenging to burn otherwise. (4) Gasification has the ability to lower emissions of greenhouse gases and other pollutants per unit of energy produced [30].

3.2 Gasification of waste

Gasification of waste has been chosen over the incineration because it provides different syngas product that utilized in numerous forms while contrasting to the hot gases of combustion. It provides the even quality synthetic gas from the varied and composite waste residues. The gasification of waste is lone option that could offer multimodal products, for example, cooling, heat, power, liquid, and gaseous fuels along with chemicals [29].

It comprises four different phases, which are following such as feedstock drying, pyrolysis, oxidation, and the reduction [35].

DOI: http://dx.doi.org/10.5772/intechopen.1002019

3.2.1 Drying

Waste feedstocks having variable moisture level have been dry in the drying process beyond 100°C. In that stage, chemical reactions do not take place, the phase alter among liquid to water and the vapor that is core cause of energy requirements in the drying procedure.

3.2.2 Pyrolysis

In pyrolysis, feedstock decomposition starts in the lack of the oxidant on high temperature, and the vapors have been free from dried feedstock by primary reactions. Proportion of vapors and the produced char are predisposed through the conditions of process, for example, temperature and heating rate. In addition, product distribution has been affected by the composition and size of feedstock [36].

3.2.3 Oxidation

On high temperature and in the environment that is partially oxidized, the heterogeneous reactions occur among CO, oxidant, and water vapor of feedstock. Oxidation reaction has influenced through chemically feedstock composition, oxidant types (such as oxygen, CO_2, air, or steam,), and the operating conditions. Generally, oxidation phase is an exothermic phase, and heat energy releases as a result of it for energy independence to endure heating requirements of the process.

3.2.4 Reduction

It is an endothermic phase through which the high-temperature chemical reaction takes place in oxygen absence. Several reactions among products of oxidation and the char occur to produce novel hydrocarbons. Some char and ash are the by-products of reduction phase [35, 37].

Advantages of gasification of waste are: waste is gasified, which minimizes the demand for landfill space, lowers methane emissions, and uses fewer fossil fuels. Thus, waste gasification is employed to improve recycling initiatives [36].

3.3 Gasification of coal

Coal gasification is a method in which coal, char, or coke is converted into gassy fuels through contact with the mediators, for example, steam, oxygen, or air. While a certain gas amount has been freed from coal through carbonization methods (such as heating of in nonexistence of oxygen or air), and complete coal gasification comprises the conversions of all carbonous material to the gassy products through reaction that are given below:

3.3.1 Primary reactions

Water gas reaction:

$$C + H_2O \rightarrow CO + H_2 \tag{6}$$

Boudouard reaction:

$$C + CO_2 \rightarrow 2CO \quad (7)$$

Partial combustion:

$$C + O_2 \rightarrow CO \quad (8)$$

Hydrogasification:

$$C + 2H_2 \rightarrow CH_4 \quad (9)$$

3.3.2 Secondary reactions

Shift reaction:

$$CO + H_2O \rightarrow H_2 + CO_2 \quad (10)$$

Methanation:

$$3H_2 + CO_2 \rightarrow CH_4 + H_2O \quad (11)$$

The only residues of the complete coal gasification would, hence, be ash-forming components of mineral matter that are released solid form (slag) [38]. Depending on coal nature and actual gasification method, products of the gasification might include CO_2, CO, CH_4, and H_2 diluted *via* H_2S and the atmospheric nitrogen. Numerous impurities mainly pollutants, for example, H_2S or noncombustible materials e.g., N_2 and CO_2 might be removed through following purification processes. Modern day, the coal gasification has been directed about manufacture of the two kinds of products:

- Synthesis gas, combination of CO and H_2, with 9 kJ m^{-3} specific energy.
- Synthetic natural gas, essentially made of methane, with 30 kJ m^{-3} specific energy.

Syngas could be utilized to produce amount of products, containing methanol, ammonia, and liquid fuels, but the synthetic natural gas is utilized as substitute for the natural gas in industrial and domestic reticulation schemes [39].

Gasifiers are used in coal gasification and are fundamentally bounded pressure vessel in which pulverized or crushed coals has been interacted with the gasifying agent. Lurgi fixed bed gasifier with the jacketed vessel of water having diameter of 4 m is one of the widely used gasifiers for coal gasification. Indelicately, the crushed coal has been fed in by lock hopper structure at the top and passes downward slowly alongside upward flow of steam and oxygen. Products of hot gasification have been removed from the top, whereas the ash has been removed innextlock arrangement on base.

A more eco-friendly and less-polluting way to treat coal is through coal gasification. When burned, it releases less carbon, and other harmful gases, such as CO2, are simple to separate, capture, and utilize in other ways. Coal gasification syngas can also be further processed to produce fuels such as petrol and diesel. Other coal gasification by-products, such as hydrogen, can be employed in the production of ammonia, the search for alternative fuel sources, or the petrochemical sector [38].

3.4 Gasification of natural gas

Partial oxidation (POx) is also known as gasification. By using natural feedstock, we use partial oxidation as gasification process to produce syngas. Chemistry of partial oxidation process has been founded on the partial fuel combustion, such as in CH_4 case, that is represented in the following equation

$$CH_4 + \frac{1}{2} O_2 = CO + 2H_2 \tag{12}$$

Though, this process of partial oxidation is mostly utilized to produce syngas from the heavy hydrocarbons, with petroleum coke and deasphalted pitch. After ignition, feedstock, such as natural gas, is preheated and then mixed with the oxygen inside burner; reaction occurred inside the elevated temperature combustion chamber produce effluent, which comprise of numerous quantities of soot, depending on composition of feedstock.

Typically, exodus gas temperature of reactors is included among 1200–1400°C. Attained synthesis gas must cool and cleaned inside "washing" sector to remove impurities. Elevated temperature (such as 1400–1100°C),the recovery of heat in partial oxidation has not been very effectual, and POx benefit over stream reforming is in option of utilizing the "low-worth" feedstocks and comprising sulfur and the other numerous compounds, which are fatal to catalysts of stream reforming. Presently, key utilizations of the partial oxidation are: (1) in the H_2 production for the refinery applications, (2) production of syngas from natural gas and (3) in electricity production from deasphalted bottoms and petroleum coke, by using large integrated gasification combined cycles (IGCC) [40].

3.5 Gasification using different gasifiers

It is the complicated process that has been influenced by numerous factors among which design of the equipment plays a significant role. Some important and popular gasifiers have been listed and briefly discoursed as bellow.

3.6 Fixed bed gasifier

It is a type of gasifier in which there are three ways to enter the gasifying agents in reactor to react with each other such as downdraft, updraft, and cross-draft gasifier. Updraft gasifier is explained below.

Updraft gasifier is also known as countercurrent gasifier in which air/oxygen has been passed through bottom level of gasifier, and produced product gases have been left at the top of gasifier [41]. The combustion reactions have been occurring at the bottom of gasifier near grate. After combustion, reduction reactions have been occurring at slightly upper level of the combustion zone, as displayed in **Figure 2**. In the upper level of the gasifier, pyrolysis and heating of feedstock materials have been occurred consuming forced convection and the radiation heat transfer method, wherever required heat has been provided from the reduction and combustion zone of

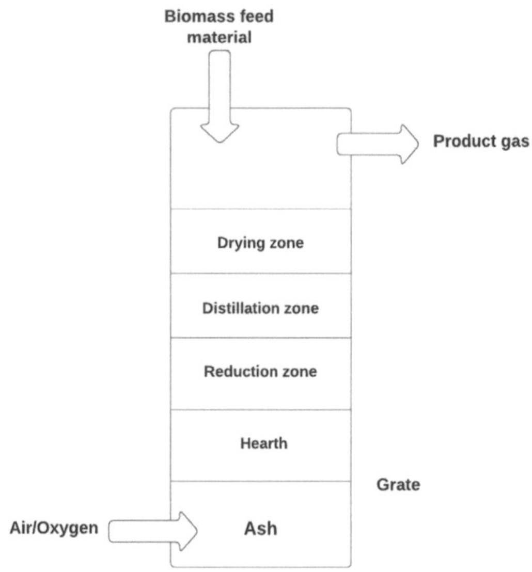

Figure 2.
Gasification process using fixed-bed (downdraft) gasifier [42].

gasifier [43]. The produced tars and volatile matters in updraft gasifier have been carried in upper-level gas stream, as showed in **Figure 2**. Alternatively, produced ash has been required to clean in bottom of updraft gasifier.

3.7 Fluidized bed gasifier

In fluidized bed gasifier, the fuel has been fluidized with air and steam. The fuel has been fed into circulating or bubbling form of fluidized bed. In fluidized bed gasifier, bed acts as fluid having high turmoil. In this gasifier, ash has been removed from gasifier in the dry state, which de-fluidize. In fluidized bed gasifier, the temperature is low, and highly reactive fuel is required [44]. Though, energy conversion efficacy has been lower than fixed bed gasifier due to elutriation of the carbonaceous fuel [45]. Fluidized bed gasifier is of three types such as bubbling, circulating, and dual fluidized bed.

The principle of downdraft and updraft and gasifier has been affected by the physical and chemical properties of fuel. Fluidized bed gasifiers could solve few problems of fixed bed gasifiers, for example, low bunker flow and pressure drop over gasifier. It has ability to tolerate high distinction of the fuel quality along with large particle distribution [46].

3.8 Entrained flow gasifier

In entrained flow gasifier, the liquid fuel, dried solid milled, or slurry has been reacted with air/oxygen in the gasification process through cocurrent flow. In this gasifier, the gasification reactions have been taken in denser cloud of the fine particles. In which high output could be attained, but overall efficacy is comparatively low than fixed-bed and fluidized bed gasifier. The residence time of the entrained flow

Gasifier	Advantages	Disadvantages
Updraft fixed bed	• Good for small-scale thermal applications • No carbon in the ash • Small pressure drop • Tolerant of high moisture	• High tar yield • Slagging potential • Feed size limit • Gas with a low heating value • Scale limitations
Downdraft fixed bed	• Low tar • Low particulates • Small scale application • Flexible adaptation	• Moisture sensitive • Gas with low heating value • Feed size limit • Scale limitations • Tall design
Fluidized bed	• High mass transfer • High heat rate • Easily scaled to huge dimensions for the production of electricity • Fuel flexibility	• Costly • Higher power • High particulate in gas • Medium tar yield
Entrained flow	• Possibility of low tar • Can be scaled • Possibility of low methane • Can generate gas with a higher heating value.	• An abundance of carrier gas • Increased particle load

Table 1.
Advantages and disadvantages of different gasifier.

gasifier is about 5 seconds, which has been shorter than residence time of fixed-bed and fluidized bed gasifier. In this type of gasifier, most of the reactions are endothermic. So, high heat has been required that is supplied through the combustion of feed material and from external sources of the heat [47].

In this type of gasifier, fine amount of coal with the air has been added co-currently, so air and steam surround coal feedstocks. Usually, it works at high temperature and pressure [48]. Therefore, flow is turbulent in entrained flow gasifier. Rate of the gasification reaction and efficacy of carbon conversion is high, although production of hydrocarbons has been low. Furthermore, coal devolatilization procedure produces oil, tar, phenol, and other liquids, which could be decomposed into hydrogen (**Table 1**).

4. Applications

4.1 Gaslighting

Before, widely availability of natural gas and electric lighting, syngas was utilized for the synthesis of illuminating gas (such as coal gas) for many years. Illuminating gas used for cooking, gaslighting, and heating. Gaslighting has been a process in which artificial light production takes place through the combustion of gaseous fuel, for example, hydrogen, carbon monoxide, methane, propane, acetylene, ethylene, butane, coal gas, and natural gas, etc. Light has been generated directly from flame, to enhance the brightness of illuminating light special combination of propane or butane has been used.

4.2 Energy capacity

Unmethanized syngas generally has a lower heating value, that is. Nearly equal to 120 BTU/scf. Hybrid turbines could be proceeding by using untreated syngas, which allows for higher efficacy due to their lower operating temperature and extended part lifetime. The output power of synthesis gas-fired turbine plant can be enhanced by up to 20–25% in comparison to the similar turbine fired at the same operating temperatures as the natural gas. Nevertheless, rise in the power output has been attributed to increase in moisture level of products of combustion. Due to the greater amount of hydrogen in syngas and amplified turbine flow, both of which make a significant contribution toward turbine component overheating [49].

4.3 Diesel

Synthesis gas could be utilized in Fischer-Tropsch technique for the production of diesel. It also used to convert syngas in methane, dimethyl ether, and methanol in the catalytic processes. Mostly, methanol has been synthesized from synthesis gas. Although most of the methanol synthesis has been based on the natural gas as feedstock, coal-derived synthesis gas also used; as solid/coal, feedstocks have been utilized to produce 9% of global yield of the methanol [50].

4.4 Hydrogen production

Syngas has been used to synthesize hydrogen by the Haber process. The synthesis gas generated because of the gasification procedure comprises hydrogen and carbon monoxide, which is again reacted with the steam to separate the hydrogen.

4.5 Power generation from syngas

The primary application of syngas is generally the generation of power and heat. This can be achieved either in autonomous combined heat and the power plants or by cogeneration of gas produced in large-scale plants. Synthetic gas from pyrolysis is a fuel gas that can be used to produce electricity in a wide range of equipment, from steam cycles to gas engines and turbines. Whereas, boiler applications for steam cycles

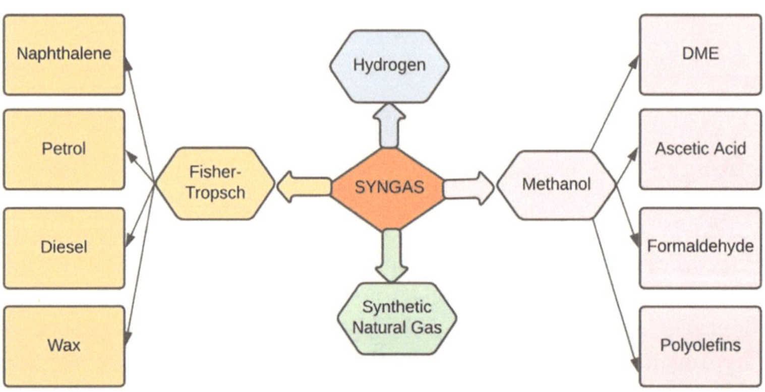

Figure 3.
Applications of syngas [52].

generally do not require intensive gas treatment prior to electricity production. Gas engines require more advanced purification and processing.

The stability and consistency of fuel provided to the internal combustion engine are one of the important elements of the technique. It is guaranteed by stability of feedstock and the precisely controlled processing conditions in the bio green pyrolysis unit [51]. Flow sheet of different applications of syngas is illustrated in **Figure 3**.

5. Conclusion

Synthesis gas is definite as gas with CO and H_2 as the key components of fuel. Gasification is one of attractive approaches to produce hydrogen-rich gas. Synthesis gas production with improved properties from different feedstocks (such as biomass, waste, coal, and natural gas) has been an attractive solution for the energy crisis. Syngas through gasification has been used in different applications to overcome the energy crises of petroleum products.

Acknowledgements

The authors acknowledge faculty of life sciences and vice chancellor for any kind of assistance provided by them.

Conflict of interest

The authors declare that there is no conflict of interest.

Author details

Saba Mahboob, Kalsoom Tahir, Sikander Ali*, Iram Liaqat and Nauman Aftab
Institute of Industrial Biotechnology (IIB), GC University Lahore, Pakistan

*Address all correspondence to: dr.sikanderali@gcu.edu.pk

References

[1] Ohlström M, Mäkinen T, Laurikko J. New concepts for biofuels in transportation biomass-based methanol production and reduced emissions in advanced vehicles. VTT Technical Research Centre of Finland. 2001;**2001**:3-94

[2] Maniatis K. In: Bridgewater AV, editor. Progress in Biomass Gasification: An Overview. Oxford: Blackwell Scientific Publications; 2001

[3] Faaij A. Modern biomass conversion technologies. Mitigation and Adaptation Strategies for Global Change. 2006;**11**: 343-375

[4] Capodaglio AG, Olsson G. Energy issues in sustainable urban wastewater management: Use, demand reduction and recovery in the urban water cycle. Sustainability. 2020;**12**(1):266. DOI: 10.3390/su12010266

[5] Martin MM. Industrial Chemical Process Analysis and Design. Cambridge, MA: Elsevier; 2016

[6] Salim W, Ho WSW. Hydrogen purification with CO_2-selective facilitated transport membranes. Current Opinion in Chemical Engineering. 2018;**21**:96-102

[7] Voldsund M, Jordal K, Anantharaman R. Hydrogen production with CO_2 capture. International Journal of Hydrogen Energy. 2016;**41**(9): 4969-4992

[8] Díaz JA, Akhavan H, Romero A, Garcia-Minguillan AM, Romero R, Giroir-Fendler A. Cobalt and iron supported on carbon nanofibers as catalysts for Fischer-Tropsch synthesis. Fuel Processing Technology. 2014;**128**:417-424

[9] Kobayashi M. Chapter 1 – Introduction. In: Dry Syngas Purification Processes for Coal Gasification Systems. Elsevier; 2021. pp. 1-49. DOI: 10.1016/B978-0-12-818866-8.00001-X

[10] Speight JG. The Chemistry and Technology of Petroleum. 5th ed. Boca Raton, Florida: CRC Press, Taylor & Francis Group; 2014

[11] Speight JG. Handbook of Petrochemical Processes. Boca Raton, Florida: CRC Press, Taylor & Francis Group; 2019

[12] Spath PL, Dayton DC. Preliminary Screening Technical and Economic Assessment of Synthesis Gas to Fuels and Chemicals with Emphasis on the Potential for Biomass Derived Syngas. Golden: National Renewable Energy Lab Golden Co.; 2003. DOI: 10.2172/1216404

[13] Roseno KT, Alves RMB, Giudici R, Schmal M. Syngas production using natural gas from the environmental point of view. Biofuels - State of Development. 2018;**2018**:97-113. DOI: 10.5772/intechopen.74487

[14] Speight JG. Shale Gas and Shale Oil Production Processes. Cambridge, MA: Elsevier Inc.; 2020

[15] Speight JG. Heavy and Extra Heavy Oil Upgrading Technologies. Oxford, United Kingdom: Gulf Professional Publishing, Elsevier; 2013

[16] Speight JG. Synthetic Fuels Handbook: Properties, Processes, and Performance. New York: McGraw-Hill; 2008

[17] Speight JG, editor. Biofuels Handbook. London, United Kingdom: Royal Society of Chemistry; 2011

[18] Speight JG. The Refinery of the Future. Oxford, United Kingdom: Gulf Professional Publishing, Elsevier; 2011

[19] Lee S, ShahYT. Biofuels and Bioenergy: Processes and Technologies. Boca Raton Florida: CRC Press, Taylor & Francis Group; 2013

[20] Hornung A. Transformation of Biomass: Theory to Practice. Chichester, West Sussex, United Kingdom: John Wiley & Sons Inc.; 2014

[21] Sheth PN, Babu B. Experimental studies on producer gas generation from wood waste in a downdraft biomass gasifier. Bioresource Technology. 2009; **100**:3127-3133

[22] Kirubakaran V, Sivaramakrishnan V, Nalini R, Sekar T, Premalatha M, Subramanian P. A review on gasification of biomass. Renew Sustainable Energy Review. 2009;**13**:179-186

[23] Zabaniotou AA, Skoulou VK, Mertzis DP, Koufodimos GS, Samaras ZC. Mobile gasification units for sustainable electricity production in rural areas: The smart-CHP project. Industrial and Engineering Chemistry Research. 2011;**50**:602-608

[24] Asadullah M. Barriers of commercial power generation using biomass gasification gas: A review. Renew Sustainable Energy Review. 2014;**29**: 201-215

[25] Radwan AM, Hamad MA, Singedy AM. Synthesis gas production from catalytic gasification of saw dust. Life Science Journal. 2015;**12**: 104-118

[26] Zhang W. Automotive fuels from biomass via gasification. Fuel Processing Technology. 2010;**91**:866-876

[27] Bhavanam A, Sastry R. Biomass gasification processes in downdraft fixed bed reactors: A review. International Journal of Chemical Engineering Applications. 2011;**2**:425-433

[28] Saghir M, Rehan M, Nizami AS. Recent Trends in Gasification Based Waste-to-Energy. Gasification for Low-grade Feedstock. Vol. 2018. London, UK: Intechopen; 2018. pp. 97-113

[29] Prins MJ, Ptasinski KJ, Janssen FJ. From coal to biomass gasification: Comparison of thermodynamic efficiency. Energy. 2007;**32**(7): 1248-1259

[30] Baliban RC, Elia JA, Floudas CA. Toward novel hybrid biomass, coal, and natural gas processes for satisfying current transportation fuel demands, 1: Process alternatives, gasification modeling, process simulation, and economic analysis. Industrial & Engineering Chemistry Research. 2010; **49**(16):7343-7370

[31] Modell M, Reid RC, Amin SI. Gasification process. 1978. Google Patents

[32] Buragohain B, Mahanta P, Moholkar VS. Biomass gasification for decentralized power generation: The Indian perspective. Renew Sustainable Energy Reviews. 2010;**14**:73-92

[33] Tavasoli A, Ahangari MG, Soni C, Dalai AK. Production of hydrogen and syngas via gasification of the corn and wheat dry distiller grains (DDGS) in a fixed-bed micro reactor. Fuel Processing Technology. 2009;**90**:472-482

[34] Das BK, Hoque S. Assessment of the potential of biomass gasification for electricity generation in Bangladesh. Journal of Renewable Energy. 2014; **2014**:10

[35] Demirbas A. Biomass gasification for power generation in Turkey. Energy Sources, Part A: Recovery, Utilization, and Environmental Effects. 2006;**28**(5): 433-445

[36] Bridgwater AV. Renewable fuels and chemicals by thermal processing of biomass. Chemical Engineering Journal. 2003;**91**(2):87-102

[37] André RN. Fluidised bed co-gasification of coal and olive oil industry wastes. Fuel. 2005;**84**(12):1635-1644

[38] Chadwick MJ, Highton NH, Lindman N. Coal conversion technologies. In: Environmental Impacts of Coal Mining & Utilization. Stockholm, Sweden: Pergamon Press, Elsevier Inc.; 1987. pp. 129-130

[39] Ward CR. Coal Geology. Netherland: Elsevier Inc.; 2013. p. 29

[40] Laquaniello G, Iaquaniello G, Antonetti E, Cucchiella B, Palo E, Salladini A, et al. Natural Gas Catalytic Partial Oxidation: A Way to Syngas and Bulk Chemicals Production. London, UK: IntechOpen; 2012. DOI: 10.5772/48708

[41] Pedroso DT, Machín EB, Silveira JL, Nemoto Y. Experimental study of bottom feed updraft gasifier. Renewable Energy. 2013;**57**:311-316

[42] Phillips J. Different types of gasifiers and their integration with gas turbines. In: Proceedings of 1. World Conference and Exhibition on Biomass for Energy and Industry. 2006. p. 1

[43] Brandt P, Henriksen UB. Decomposition of tar in gas from updraft gasifier by thermal cracking. In: 1st World Conference and Exhibition on Biomass for Energy and Industry. USA: Natural Energy Technology Department; 2000

[44] Mansaray K, Ghaly A, Al-Taweel A, Hamdullahpur F, Ugursal V. Air gasification of rice husk in a dual distributor type fluidized bed gasifier. Biomass and Bioenergy. 1999;**17**(4): 315-332

[45] Basu P. Combustion and Gasification in Fluidized Beds. Boca Raton: CRC Press; 2006

[46] Warnecke R. Gasification of biomass: Comparison of fixed bed and fluidized bed gasifier. Biomass and Bioenergy. 2000;**18**(6):489-497

[47] Wen CY, Chaung T. Entrainment coal gasification modeling. Industrial & Engineering Chemistry Process Design and Development. 1979;**18**(4):684-695

[48] Kajitani S, Suzuki N, Ashizawa M, Hara S. CO_2 gasification rate analysis of coal char in entrained flow coal gasifier. Fuel. 2006;**85**(2):163-169

[49] Oluyede EO, Phillips JN. Fundamental impact of firing syngas in gas turbines. In: Proceedings of the ASME Turbo Expo 2007: Power for Land, Sea, and Air. Vol. 3. ASME; 2007. pp. 175-182. DOI: 10.1115/GT2007-27385

[50] State of the Gasification Industry. Worldwide Gasification Database 2014 Update Gasification Technologies Conference. Washington, DC: Chris Higman; 2014

[51] Chatterjee A. Sponge Iron Production by Direct Reduction of Iron Oxide. United States: PHI Learning; 2012. ISBN 978-81-203-4659-8. OCLC 1075942093

[52] Jie Y, Odriozola JA, Reina TR. Dry reforming of ethanol and glycerol: Mini-review. Journal of Catalysis. 2019;**9**:1015. DOI: 10.3390catal9121015

Chapter 3

Study of the Heterogeneous Catalytic Model for Syngas Production in a Simulated Environment for Application in CO_2 Capture Process

Martha Siles Camacho and Valeria Sandoval Cossío

Abstract

The study was carried out under the possibility of transforming conventional petrochemical processes, such as ammonia production plant, into carbon dioxide capture systems by means of replacing fossil natural gas with biogas, generated by manure during an anaerobic digestion. Three partial substitution scenarios were proposed: 30, 40 and 50% in order to analyse the behaviour of CO_2 in the syngas unit process, applying an equivalent CO_2 balance to evaluate if it is possible to capture this GHG within the same process, thus achieving negative emissions. The simulation of heterogeneous catalytic model of the methane reforming reactions, allowed us to assess the kinetic behaviour of all reactions involved and modify operating conditions in order to increase the production of CO_2 and maintain its required flow as a secondary product, and at the same time to decrease CO_2 emissions. The result shows that a 30% substitution is possible in order to achieve a −1.48e6 lb/h of equivalent CO_2.

Keywords: syngas, heterogeneous catalysis reactions, capture of CO_2, simulation process, biogas, ammonia

1. Introduction

Methane was the major component of the early atmosphere of the planet and a basic raw material for the petrochemical industry, but its global warming potential GWP is 27–30 over 100 years more than that of carbon dioxide CO_2, which is why it constitutes a GHG thus contributing the greatest to global warming [1].

This chapter aims to assess the process production of syngas for an ammonia plant, with the application of a simulation model to the kinetic water steam reforming of natural gas, based on the heterogeneous catalyst of the chemical reactions involved, which are developed by [2], but also using a partial substitution of raw material studying scenarios with a mixture of biogas produced from animal manure to the process.

This study could also consider CO_2 capture due to the decrease of fossil feedstock and the substitution for biogas to assess CO_2 eq emissions generated in the process to the atmosphere.

The operative conditions with influence on the reaction rates of the process will be adjusted to maintain the original balance of CO_2 that is required by the urea plant and analyse the presence of higher concentrations of CO_2 in the process (CO_2 capture) and about the overall kinetics reactions.

The concepts of "carbon capture" and "global carbon capture" refer to the efficiency of conversion of a carbon source, such as a raw material, into products. For example, the amount of carbon in a wood biomass feedstock that is converted into useful products, such as alcohol [3]. For this case study, the partial substitution of fossil natural gas with biogas can convert the syngas unit process into a carbon capture process.

The application of the one-dimensional heterogeneous mathematical models developed by Sandoval [2] is the tool to study the heterogeneous catalyst behaviour within the water steam reformers (SMR) inside an ammonia plant. The calculations and analysis integrate the reaction mechanism, reaction rates, complex stoichiometry as a function of partial pressures, catalyst data, reactor geometry, packed bed reactor design equations, Ergun equation for the pressure drop, and the set of equations that integrate the energy balance [2].

The analysis should focus on the influence of increased CO_2 concentration on feed streams, how reaction balances can be changed within SMR reactors, and the molar ratio of reactants and products by manipulating operating conditions and process severity. Finally, the analysis of the avoided trillions of standard cubic feet fossil methane decrease and its CO_2 equivalence in emissions.

2. Characterisation of the raw material and scenarios definition for partial substitution with biogas

2.1 Natural gas

Conventional production of syngas comes from a variety of raw materials ranging from natural gas to coal. The choice for a particular raw material depends on the cost and availability of the feedstock, and on the downstream use of syngas. Syngas is generally produced by one of three processes, which are distinguished based on the feedstock used [4]:

- Steam reforming of natural gas or light hydrocarbons, optionally in the presence of oxygen or carbon dioxide.
- Partial oxidation of (heavy) hydrocarbons with steam and oxygen.
- Partial oxidation of coal (gasification) with steam and oxygen.

For this study, we will use a Bolivian natural gas composition that is considered a dry gas due to its methane concentration being greater than 88% by weight. One of its main benefits is that it has no hydrogen sulphide (H_2S) and other contaminants. **Table 1** indicates the natural gas chromatography at the exit of four primary processing plants existing in Bolivia. The composition of Carrasco will be the feedstock of natural gas for the study.

Component	Rio Grande	Yapacani	Carrasco	Vuelta Grande
N_2	1.85	1.7	0.33	1.79
CO_2	0.94	1.87	1.94	0.06
C_1	92.32	91.9	91.06	88.01
C_2	4.5	2.55	6.45	9.14
C_3	0.35	0.99	0.21	0.93
iC_4	0	0.17	0	0.03
nC_4	0.01	0.34	0	0.03
iC_5	0.01	0.12	0	0.01
nC_5	0.01	0.12	0	0
C6	0	0.12	0	0
C7+	0.02	0.13	0.01	0

Table 1.
Natural gas chromatography of selected plants [2].

2.2 Substitute raw material (biogas)

Technological advances and growing concern about global warming due to greenhouse gas emissions (GHG) have led to finding new raw materials that replace fossil resources to generate energy and reduce emissions in industrial processes.

Biomass is an alternative that is taking its place in technological processes and, at the same time, to use organic waste. **Figure 1** provides a detailed illustration of the many potential biomass-to-energy technologies and industrial feedstocks, which are at varying technology readiness levels (TRLs). Our route is highlighted in red [6].

Biochemical pathways take advantage of anaerobic digestion and fermentation to produce fuels, ex. biomethane [6].

Biogas or biomethane is a biofuel produced by means of (i) organism anaerobic digestion, which digests material inside a closed system or (ii) fermentation of biodegradable organic matter including manure, sewage sludge, municipal solid waste, biodegradable waste, or any other biodegradable feedstock under anaerobic conditions. The typically anaerobic process of biogas production is a multistep biological process where the originally complex and big-sized organic solid wastes are progressively transformed into simpler and smaller-sized organic compounds by different bacterial strains to have a final energetically worthwhile gaseous product and a semi-solid material (digestate), rich in nutrients and thus suitable for use in farming [7].

While natural gas is classified as a fossil fuel, biomethane is defined as a non-fossil fuel [8], and it is further characterised or described as a green energy source. It is noteworthy that methane anyway, whatever the source (fossil fuel or non-fossil fuel), when released into the atmosphere, has 20 more GWP as a greenhouse gas than that of CO_2. Organic matter from which biomethane is produced would release the CO_2 into the atmosphere if simply left to decompose naturally, while other gases that are produced during the decomposition process, for example, nitrogen oxide(s), would make an additional contribution to the greenhouse effect [7].

Biological biomethane, as well as other biofuels, can be used to generate heat and energy or also used as raw material in petrochemical transformation processes such as the production of ammonia. We will use it as a secondary raw material (partial

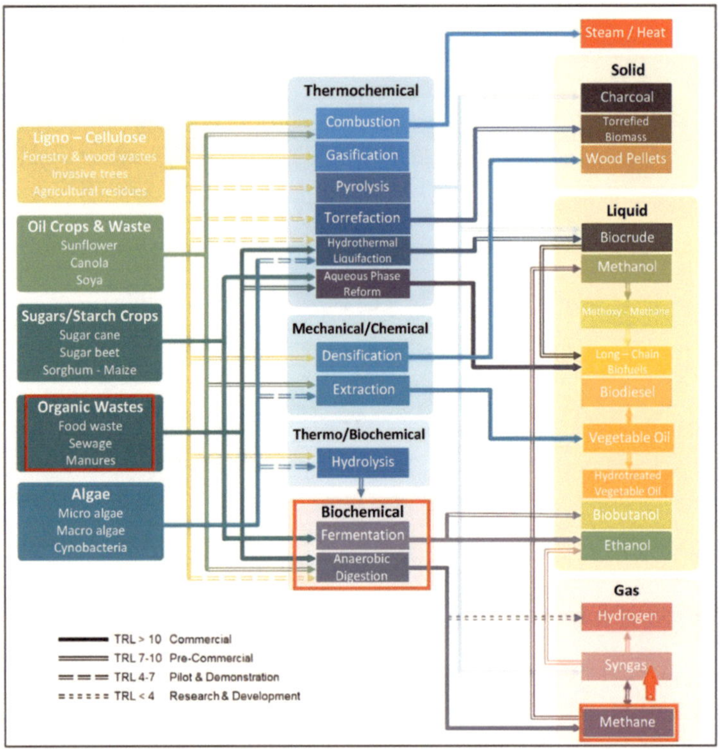

Figure 1.
Biomass conversion pathways and technology readiness levels (TRLs). Adapted from [5].

substitute). Biomethane is also called biogas. The data came from a production cycle by anaerobic digestion with manure of animal origin.

Intensive pig production generates considerable quantities of manure, and its final disposal represents a major environmental problem. Uncontrolled applications of pig manure over oil can cause an excess of nitrates, salts, heavy metals (copper and zinc), pathogens, xenobiotic compounds, and emissions of GHG [9].

Pigs produce 4 kg of fresh manure per day per head, so each pig can produce 0.33 m^3 of useful biogas per day [9].

In this case, a composition of biogas produced from pig manure will be used. The data were acquired from the biogas experimental unit at the Universidad Privada del Valle (Bolivia). **Table 2** shows the concentration measured after the contaminant's removal process in batch production:

At this point, it is important to note that the contaminants of the biogas, mainly hydrogen sulphide, are fully removed after the digestion process. Therefore, the secondary feedstock that will serve as a substitute for fossil natural gas is free of this significant contaminant for the process.

2.3 Process flow analysis for the study

The natural gas is mixed with steam and enters the tubes of the radiant zone of the primary reformer (PR), where both the steam reforming reactions and the water gas shift (WGS) reactions take place [2].

Batch	L-1803
Compound	
Methane (CH_4) %	60.66
Carbon dioxide (CO_2) %	39.3
Oxygen (O_2) %	0.0003
Hydrogen sulphide (H_2S) ppm	0
Carbon monoxide (CO) ppm	15

Table 2.
Results of the biogas analysis, adapted from [10].

The process gas enters the secondary reformer (SR), where the amount of air is not controlled in order to produce a molar ratio of 3:1 hydrogen/nitrogen. As a matter of fact, extra air is used to provide additional heat to the SR, and there is a leak of 1.75% methane on a dry basis. During secondary reforming, the oxygen produces the combustion of the partial flow of methane, and then in the catalytic zone the reforming reactions will occur without heat transfer [2].

The WGS reaction occurs for the CO to CO_2 conversion. This conversion occurs in two stages: in the high-temperature converter (HTS) and the low-temperature converter (LTS). The effluent gas is directed towards the CO_2 absorption unit [2].

Within the industrial process of ammonia, there are several products and services that are part of its life cycle assessment (LCA), which are both directly and indirectly connected. The complete plant can be considered as a system whose boundaries are many interrelated processes that constitute the basic components of the system or "unitary processes".

The reforming unit represents the main unitary process of the system for the study (**Figure 2**), from which we will calculate the input and output flows. In addition, our objective is to analyse the system as a potential CO_2 capture process within the life cycle of ammonia production.

The following diagram indicates the proposed CO_2 elimination route, in which we can observe that the secondary source reduces the flow of fossil natural gas feed to the process (**Figure 3**).

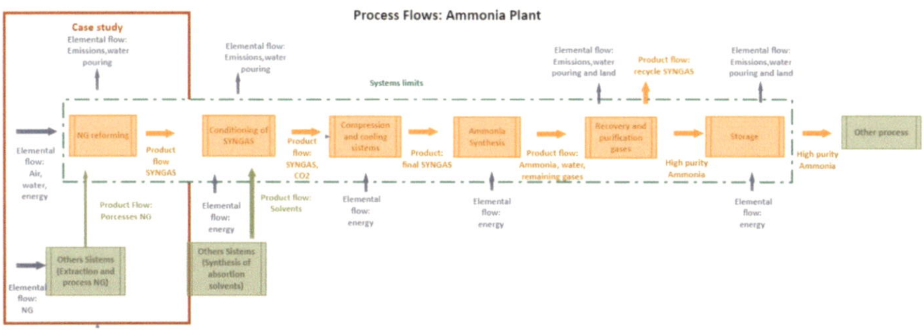

Figure 2.
Process flow inside the ammonia plant. Own elaboration.

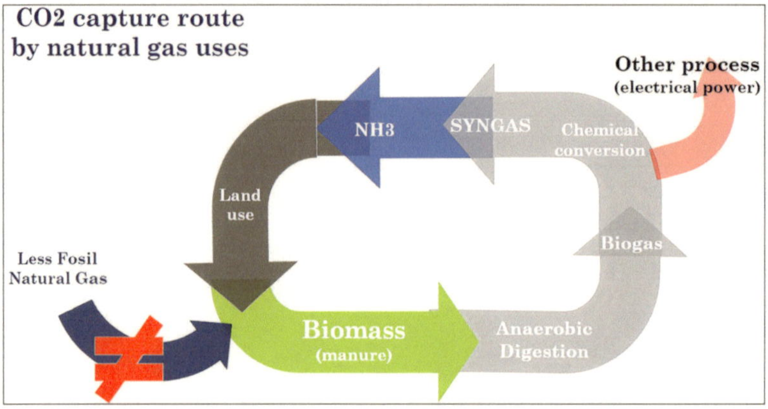

Figure 3.
Proposed CO_2 removal route. Own elaboration.

2.4 Fossil natural gas substitution scenarios

To determine the scenarios of replacing natural gas with biogas, we will use set points to monitor the variation of the feed flow, the conversion and the final balance regarding the CO_2 in the syngas unit.

Feed process composition by scenario					
	Carrasco	Basic design	S1 (70–30)	S2 (60–40)	S3 (50–50)
Component mole fraction	GN	GN*	GN-BG	GN-BG	GN-BG
Methane	0.9106	0.8864	0.8193	0.7935	0.7655
CO_2	0.0194	0.0189	0.0979	0.1284	0.1613
Ethane	0.0645	0.0628	0.0490	0.0436	0.0378
Propane	0.0021	0.0020	0.0016	0.0014	0.0012
i-Butane	0.0000	0.0000	0.0000	0.0000	0.0000
n-Butane	0.0000	0.0000	0.0000	0.0000	0.0000
n-Pentane	0.0000	0.0000	0.0000	0.0000	0.0000
i-Pentane	0.0000	0.0000	0.0000	0.0000	0.0000
n-Hexane	0.0000	0.0000	0.0000	0.0000	0.0000
n-Heptane	0.0001	0.0001	0.0001	0.0001	0.0000
Nitrogen	0.0033	0.0098	0.0098	0.0098	0.0099
Hydrogen**	0.0000	0.0200	0.0222	0.0231	0.0241
Molar flow (lbmol/h)	5363.96	3524.00	3164.56	3044.75	2924.94
Mass flow (lb/h)***	94382.12	61152.98	61152.98	61152.98	61152.98

**The study considers the substitution by biogas only for the natural gas in the proce but not the fuel for the furnace or other facilities. **A small fraction of hydrogen injection is considered due to the indeterminacies caused by its absence in the reaction rates [11]. ***The total mass flow of feedstock to the process remains constant.*

Table 3.
Substitution scenarios of natural gas by biogas.

The percentage of substituted natural gas and the global composition in the feed will be the basis for defining the scenarios in this study. By manipulating the operating conditions in the reactors without exceeding the severity of the process, we will start from the hypothesis that the kinetics of the reforming reactions can be controlled to accomplish the balance towards the formation of syngas with a 3:1 hydrogen-nitrogen ratio and to maintain the balance of CO_2 used as a secondary product (urea). The avoided CO_2 indicator is considered from two points of view:

- The decrease of fossil natural as a raw material.
- Biogas feeding from organic matter residue whose emissions would be avoided.

Table 3 shows the substitution scenarios proposed with the secondary raw material.

Table 3 shows the substitution scenarios of natural gas by biogas for the study, considering three mixes in percentage: S1 (70% GN 30% BG), S2 (60% GN 40% BG) and S3 (50% GN 50% BG). We can notice that the composition of CO_2 is increasing from 1.89% in the basic design to 16.13% in the S3 scenario.

3. Heterogeneous catalytic model of natural gas reforming reactions in the syngas process

Reactions 1 and 3 are the steam reforming reactions for methane and Reaction 2 is the water gas shift reaction,which takes place simultaneously. The water gas shift reaction is fast and is generally considered in equilibrium [12]

$$CH_4 + H_2\,O \leftrightarrow 3H_2 + CO \quad (1)$$

$$CO + H_2O \leftrightarrow CO_2 + \ H_2 \quad (2)$$

$$CH_4 + 2\,H_2O \leftrightarrow CO_2 + 4\,H_2 \quad (3)$$

"Hinshelwood" mechanism for the three steam reforming reactions is:

$$H_2O + Sc \leftrightarrow O - Sa + H_2 \quad (4)$$

$$CH_4 + Sc \leftrightarrow CH_4 - Sa \quad (5)$$

$$CH_4 - Sa + Sc \leftrightarrow CH_3 - Sa + H - Sa \quad (6)$$

$$CH_3 - Sa + Sc \leftrightarrow CH_2 - Sa + H - Sa \quad (7)$$

$$CH_2 - Sa + O - Sa \leftrightarrow CH_2O - Sa + Sc \quad (8)$$

$$CH_2O - Sa + Sc \leftrightarrow CH_2O - Sa + H - Sa \quad (9)$$

$$CHO - Sa + Sc \leftrightarrow CO - Sa + H - Sa.\text{controlling Step Reaction 1} \quad (10)$$

$$CO - Sa + O - Sa \leftrightarrow CO_2 - Sa + Sc \text{ controlling Step Reaction 2} \quad (11)$$

$$CHO - Sa + O - Sa \leftrightarrow CO_2 - Sa + H - Sa \text{ controlling Step Reaction 3} \quad (12)$$

$$CO - Sa \leftrightarrow CO + Sc \quad (13)$$

$$CO_2 - Sa \leftrightarrow CO_2 + Sc \quad (14)$$

$$2H - Sa \leftrightarrow Sc + H_2\text{–}Sa \quad (15)$$

$$H_2 - Sa \leftrightarrow H_2 + Sc \quad (16)$$

Sc is the catalyst surface; Sa is the adsorbed catalyst.

	Preexponential factor: A (k_i)	Activation energy and E_i (kJ/mol)
$k1\_$	4225e15	240.1
$k2\_$	1995e6	67.13
$k3\_$	1020e15	243.9
K_{CH4} bar $^{-1}$	6.65e−4	−38.28
KH2O_	1.77e5	88.68
KCO bar $^{-1}$_	8.23e−5	−70.65
K_{H2} bar^{-1}	6.12e−9	−82.9
K_1	1.19136e13	−223.02
K_2	1.6e−2	36,565
K_3	1.957e11	−186,464

Table 4.
Kinetic constants of steam reforming reactions, adapted from [2, 13, 14].

The reaction rates, Eqs (17)-(20), are a function of the partial pressures per component and the constants of reaction rate, reaction equilibrium and adsorption per component. They cannot be used when the hydrogen concentration is null because they predict indeterminacies (**Table 4**) [14].

$$r_1 = \frac{k_1\left(\frac{P_{CH4}\,P_{H2O}}{P_{H2}^{2.5}} - \frac{P_{CO}\,P_{H2}^{0.5}}{K_1}\right)}{DEN^2} \tag{17}$$

$$r_2 = \frac{k_2\left(\frac{P_{CO}\,P_{H2O}}{P_{H2}} - \frac{P_{CO2}}{K_2}\right)}{DEN^2} \tag{18}$$

$$r_3 = \frac{k_3\left(\frac{P_{CH4}\,P_{H2O}^{2}}{P_{H2}^{3.5}} - \frac{P_{CO2}\,P_{H2}^{0.5}}{K_1K_2}\right)}{DEN^2} \tag{19}$$

$$DEN = 1 + K_{CH4}\,P_{CH4} + \frac{K_{H2O}\,P_{H2O}}{P_{H2}} + K_{H2}\,P_{H2} + K_{CO}\,P_{CO} \tag{20}$$

The study of the operation of natural gas reforming reactors with steam and air in order to model the catalytic kinetics of a one-dimensional heterogeneous reaction model (with the catalyst assuming state variables as a function of the length of the reactor as axial coordinate) integrates the following considerations:

- The system operates in steady state.
- Hydrocarbons higher than methane are considered inert since the reaction system inside the reforming tube is described by the three kinetic expressions of Xu and Froment.
- Assumes use of fresh and activated Haldor Topsoe catalyst R67-7H.
- The stoichiometry will be related to the disappearance of methane and the production of CO_2.

3.1 Simultaneous kinetics reactions

In complex reactions, stoichiometry requires the variables of the consumption and production of compounds as stated by each reaction. In this case, they will be a function of the production of CO, CO_2 and hydrogen with respect to Reactions 1 and 3 (represented by x and z respectively). Reaction 2 (y) is left out of the total sum of fluxes, but it is intrinsic in equilibrium:

$$\frac{F_T}{F_{T0}} = 1 + \varepsilon\,(X + Z) \qquad \varepsilon = 2\,y_{CH40} \tag{21}$$

Since the three reactions are not independent, it is necessary to combine the three equations into two, one for the conversion of methane, and the other to produce CO_2.

$$X_{CH4} = X_A = X + Z \tag{22}$$

$$X_{CO2} = X_E = Y + Z \tag{23}$$

The behaviour of the stoichiometry of the reactions to integrate the kinetics of the reforming reactions with steam is used as a function of the partial pressures:

$$P_{CH4} = \frac{P_{CH4_0}(1 - X_{CH4})}{[1 + \varepsilon\,X_{CH4}]}\,\frac{P}{P_0} \tag{24}$$

$$P_{H2O} = \frac{P_{CH4_0}(\Theta_{H2O} - X_{CH4} - X_{CO2})}{[1 + \varepsilon\,X_{CH4}]}\,\frac{P}{P_0} \tag{25}$$

$$P_{CO} = \frac{P_{CH4_0}(\Theta_{CO} + X_{CH4} - X_{CO2})}{[1 + \varepsilon\,X_{CH4}]}\,\frac{P}{P_0} \tag{26}$$

$$P_{H2} = \frac{P_{CH4_0}(\Theta_{H2} + 3\,X_{CH4} + X_{CO2})}{[1 + \varepsilon\,X_{CH4}]}\,\frac{P}{P_0} \tag{27}$$

$$P_{CO2} = \frac{P_{CH4_0}(\Theta_{CO2} + X_{CO2})}{[1 + \varepsilon\,X_{CH4}]}\,\frac{P}{P_0} \tag{28}$$

$$P_T = \frac{P_{CH4_0}(\Theta_T + 2X_{CH4})}{[1 + \varepsilon\,X_{CH4}]}\,\frac{P}{P_0} \tag{29}$$

**Design equations are based* on *a packed bed reactor as a function of reactor length to evaluate CH_4 and CO_2 conversion.*

For the reaction rates per component, the effectiveness factors are considered, which are polynomial equations as a function of the length of the reactor obtained from a series of tests and calculations considering different steam methane ratios that allow simplifying the design of the reactor.

Conversion for methane

$$dX_A = \int_0^z \frac{\rho_b * A * -r'_A}{F_{A_0}}\,dz \tag{30}$$

Conversion for CO_2

$$dX_E = \int_0^z \frac{\rho_b * A * r'_E}{F_{A_0}}\,dz \tag{31}$$

Reaction rates per component

$$r'_{A(CH_4)} = -n1 * r'1 - n3 * r'3 \tag{32}$$

$$r'_{B(H_2O)} = -n1 * r'1 - n2 * r'2 - n3 * 2 * r3 \tag{33}$$

$$r'_{C(CO)} = n1 * r'1 - n2 * r'2 \tag{34}$$

$$r'_{D(H_2)} = n1 * 3 * r1 + n2 * r'2 + n3 * 4 * r'3 \tag{35}$$

$$r'_{E(CO_2)} = n2 * r'2 + n3 * r'3 \tag{36}$$

(ni) is the reaction effectiveness factor.
n1 = 0.07 n2 = 0.7 n3 = 0.06

We applied the Ergun equation for pressure drop based on the factor α, which integrates the total mass flow per area of the tube, the porosity, diameter and density of the particle, the length of the tube, the density of the catalytic bed, the viscosity of the mixture, the total sum of flows, initial pressure and temperature ratio.

$$\frac{dy}{dz} = \frac{-\alpha * \rho_b * A * (1 + \varepsilon * X_{CH4})}{2y} * \frac{T}{T_0} \tag{37}$$

For the energy balance model, the equations of the heat of reaction, pressure differential, reactor area and density of the catalytic bed, as well as the total sum of flows and the sum of partial pressures per heat capacity of each component, were considered.

3.2 Secondary reformer combustion zone kinetics

The complete combustion reaction is the most representative of the model for an adiabatic reactor oriented to the production of ammonia. The reaction speed is found in the same way as a function of the partial pressures, reaction, and adsorption constants per component. The design of the reactor is a function of its length, and the total flow will be equal to the initial flow since the sum of the variables of consumption and production of compounds with respect to the reaction is zero.

$$CH_4 + 2O_2 \rightarrow CO_2 + 2H_2O \tag{38}$$

$$r4 = \frac{k1 * P_{CH4} * P_{O2}{}^{0,5}}{\left(1 + K_{CH4} * P_{CH4} + K_{O2} * P_{O2}{}^{0,5}\right)^2} \tag{39}$$

$$\frac{dfCH_4}{dz} = r4\ (rCH_4) \tag{40}$$

$$F_T = F_{CH4o} * (\theta_T + 0) \tag{41}$$

$$\text{Total flow } F_T = F_{TO} + 0 \text{ then} : F_T = F_{TO.}$$

We will consider that there is no pressure drop because the reaction is very fast. The stoichiometry is placed according to the partial pressures with (Ci = Pi/R*T); therefore, we obtain the following expressions:

$$\frac{P_{CH4}}{RT} = \frac{P_{CH4o}}{RT_0}(1 - x) * \frac{P}{P_0} * \frac{T_0}{T} \tag{42}$$

$$P_{CH4} = P_{CH4o}\,(1 - x) \tag{43}$$

$$P_{H2O} = P_{CH4o}(\theta_{H2O} + 2x) \tag{44}$$

$$P_{CO} = P_{COo} \tag{45}$$

$$P_{H2} = P_{H2o} \tag{46}$$

$$P_{CO2} = P_{CH4o}(\theta_{CO2} + x) \tag{47}$$

$$P_{O2} = P_{CH4o}(\theta_{O2} - 2x) \tag{48}$$

$$P_{I1} = P_{I1o} \tag{49}$$

$$P_{I2} = P_{I2o} \tag{50}$$

The energy balance equation integrates the same parameters as the SMR reforming equation. An important aspect that is considered in the combustion stage is that the reaction is carried out until the oxygen concentration is zero.

3.3 Criticality conditions in syngas process

3.3.1 Operating temperature in the reformers

The higher outlet temperature in the primary reformer implies a higher conversion of methane to syngas (Reaction 1) and therefore a higher heat flow, that is, more fuel gas and a reduction in the airflow in the furnace, if reforming is carried out in two steps.

The operating temperature of the secondary reformer depends on two factors: the amount of air entering the process and the severity of the conditions in the primary reformer. When excess air enters the secondary reformer, the complete combustion Reaction 38 generates enough energy for the SMR reactions to take place in the catalytic zone (CZ).

The highest CO_2 production occurs in the primary reformer and the combustion zone (H_Z) of the secondary reformer (if it is the complete combustion reaction, as in our case study). The WGS reactors are necessary to modify the balance of Reaction 2 and increase the production of CO_2 and H_2.

3.3.2 Adiabatic condition in the secondary reformer

Adiabatic secondary reformers help to consume the methane. When the reformer is not adiabatic, the temperature is controlled with enveloping cooling systems in order to quantify and regulate the remaining methane flow. This is the case for the basic design of the study, which will be subject to modifications to achieve the highest possible methane consumption.

3.3.3 Excess air

If excess air enters the secondary reforming process, there will be a large amount of inert nitrogen at the outlet of the syngas unit, which can be removed by cryogenic purification.

4. Effect of the increase CO_2 flux in the SMR reforming reactions

The study model designed in Aspen HYSYS V.14, adapted from [2], considers the kinetic rate equations for Reactions (1-3) and (38), where the simulation of the reactors is defined from

a. Operating conditions and feedstock flow to the process.

b. Simultaneous reactions that take place in each reactor and kinetic parameters.

c. Catalyst properties: equivalent diameter and sphericity of the catalyst particles, catalyst density, and heat capacity of the catalyst solid.

d. Reactor size and design parameters: volume, length, diameter, and number of tubes.

e. Reactor outlet conditions.

4.1 Reaction kinetics assessment for base design and substitution scenarios

4.1.1 Primary reformer assessment

In **Figure 4**, we can see the reaction rate in the primary reactor for a basic design. Reaction 1 shows a decreasing trend of conversion, while Reaction 3 increases its conversion rate beginning at a 5 ft length of the reactor. We can also observe that the Reaction 2 remains constant, which means that there is no conversion in any direction of the equilibrium reaction.

The behaviour of the reaction rate in scenarios S1, S2 and S3 has the same trend as the basic design, but small increases in the conversion degree are observed in **Figure 5**. The global conversions of all reactors will be analysed later.

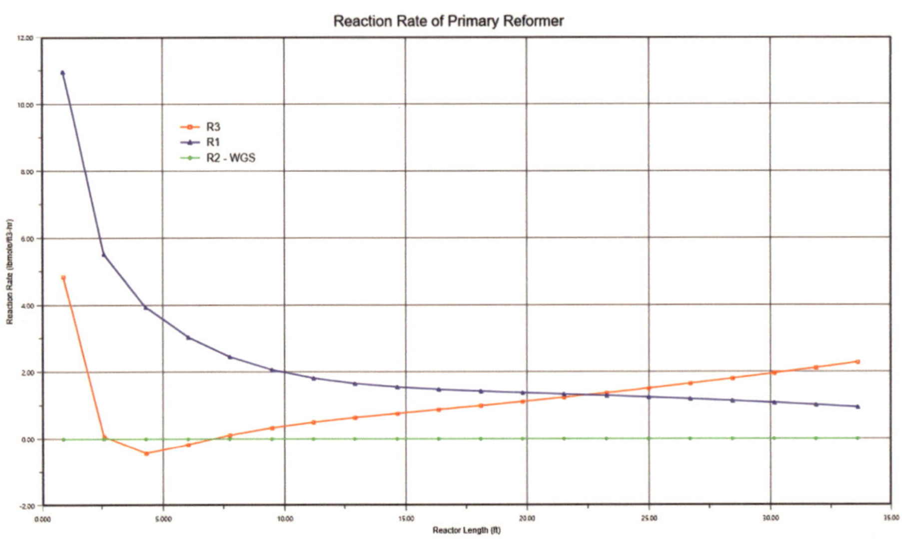

Figure 4.
Reaction rate in the primary reactor on basic scenario.

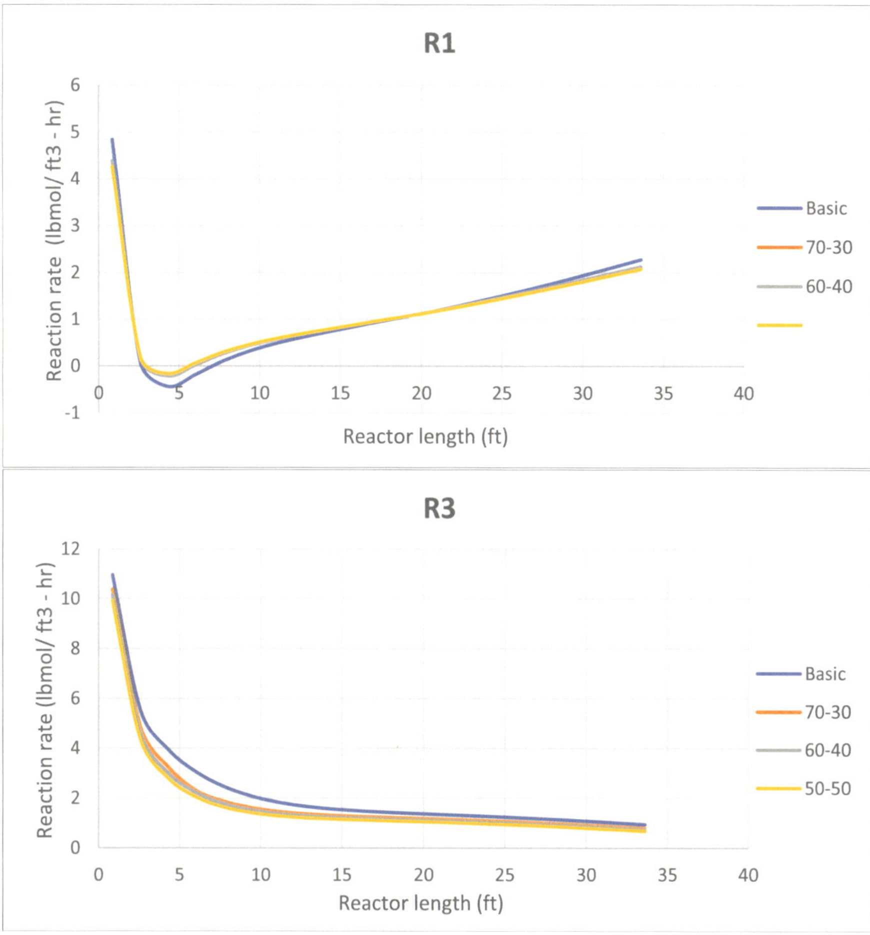

Figure 5.
*Reaction rate for basic design and scenarios S1, S2 and S3. *Reaction 2 is not included in the graph because its molar equilibrium is linear, with no apparent conversion.*

Figure 6 indicates the increased flow in the products of Reactions 1 and 3, where hydrogen has the greatest formation in quantity because it is produced in both reactions. The rising lines of hydrogen, CO, and CO_2 throughout the entire reactor indicate that Reactions 1 and 3 shift their equilibrium to the right and no reversibility occurs.

The positive conversions of methane in primary reformer let the production of both CO and CO_2 increase in the basic design and all scenarios, where CO_2 has a greater flow compared to that of CO. We will see this behaviour on CO_2 balance later.

4.1.2 Secondary reformer assessment

The designed model is a 2-step autothermal reformer (ATR), where the complete oxidation reaction takes place; this reaction is strongly exothermic, so it does not require external heat for the catalytic zone (CZ), where the SMR reactions occur again. The conversion begins with homogeneous reactions in a burner and is completed by heterogeneous catalysts [2].

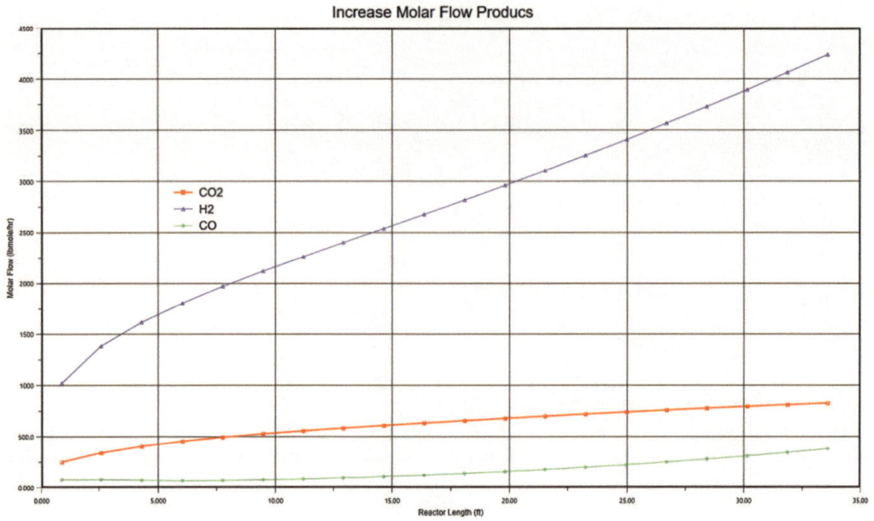

Figure 6.
Products flow increased for Reactions 1 and 3.

The kinetics of the reactions in the ATR were designed in two steps: the complete combustion zone (HZ) and then the catalytic zone CZ, where the reforming reactions interact again. The combustion of natural gas is immediate and increases the reactor temperature enough to guarantee subsequent reformation in the CZ. The outlet temperature is adjusted based on a 1.72% methane remaining (dry basis) for the basic design and all scenarios.

Figure 7 shows the component product rates in the CZ of the ATR reactor for the basic design of methane and CO_2. It should be noted that, in the beginning, the methane is consumed almost in its entirety, but throughout the reactor, it is formed again. In a similar manner, we can observe that the CO_2 production rate decreases, mainly due to

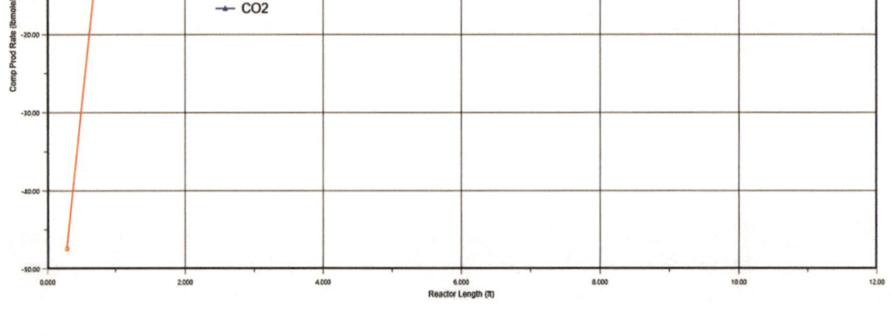

Figure 7.
Component product rate in the CZ of the ATR reactor. Basic design to methane and carbon dioxide.

the reversible behaviour of the Reaction 3, which means the reaction equilibrium shifts to the left, and CO_2 is consumed, thus forming methane and steam later, we will analyse this with the reaction conversions. Therefore, if the methane generated is available, Reaction 1 exceeds its 100% conversion from the original stream.

The kinetics of study scenarios S1, S2 and S3 present a similar behaviour, with the difference that the conversion decreases slightly as the CO_2 flux increases at each substitution percentage.

4.1.3 HTS and LTS reactors

In HTS and LTS reactors, the water gas shift reaction (WGS) takes place [15], and it is a reversible exothermic reaction between CO and steam that produces hydrogen and CO_2 (Reaction (2)).

The WGS reaction is ($\Delta H = -41.16$ kJ/mol, gas phase) and is a typical example of an equilibrium-controlled reaction, especially at higher temperatures. The reaction proceeds with no change in the number of moles, and consequently, pressure has no significant effect on equilibrium [2, 15].

In both cases, the basic design and all the scenarios, the WGS reaction moves to the right towards CO_2 and steam formation. The HTS and LTS reactors are considered the last stage to adjust the flow of CO_2 produced as a secondary product for the urea plant.

4.2 Comparative analysis of the results by scenarios

4.2.1 Steam methane reformers (SMR)

In point three, we describe the behaviour of the reforming reactions in the reactors of the syngas unit, and we also observed that the scenarios have a similar behaviour regarding the increase in CO_2. **Table 5** shows the comparative results.

Equipment		Basic design	S1 (70–30)	S2 (60–40)	S3 (50–50)
GN feed (lb/h)		94382.12	76274.66	70238.85	64203.03
1st reformer	% Reaction 1	12.63	14.83	15.84	16.95
	% Reaction 3	24.17	25.09	25.54	26.09
2nd reformer	TS HZ* (°F)	2313	2357	2372	2388
	% Reaction 38	34.53	44.05	48.29	54.5
	TS CZ** (°F)	1650	1566	1532	1495
	% Reaction 1 CZ**	118.3	107.67	103.1	97.72
	% Reaction 3 CZ**	-40.86	-36.74	-35.98	-36.08
	Methane remaining (lb/h)	4727.64	4104.38	3890.74	3676.11
	Outlet molar fraction of CH_4	0.012	0.011	0.011	0.011
HTS reactor	% Reaction 2 (HTS)	70.09	73.87	75.72	77.45
LTS reactor	% Reaction 2 (LTS)	89.58	91.99	92.2	92.64

**HZ: combustion zone.*
***CZ: catalytic zone.*
The order of the colors is decreasing, where the darker shade indicates the higher result.

Table 5.
Comparative results of operating scenarios.

The natural gas feed flow to the entire operating unit is constant, but the natural gas feed to the process decreases in proportion to the biogas substitution percentage. This is because the study considers a natural gas substitution volume only to the process; it does not consider the flows required for fuel gas from the furnace and other energy requirements in other facilities.

The primary reformer shows an increase in the conversion of Reactions 1 and 3 as the volume of substitution increases. Reaction 1 increases conversion by 17% in scenario S1 vs. the basic design and for subsequent scenarios by 7% in both cases. Reaction 3 increases 4% and 2%, respectively, in the same comparative relationship as Reaction 1. This behaviour is explained by the kinetics of methane disappearance and the amount of CO_2 present, which influences the molar balance of Reaction 3, so in the case of Reaction 1, the increased conversion of the scenarios is greater because there is no original CO in the feedstock and the molar equilibrium displacement is in greater proportion to the right compared with Reaction 3.

For the secondary reformer, a similar behaviour occurs, the combustion zone presents increasing outlet temperatures ranging from 2313°F in the basic design to 2388°F in the S3 scenario. For the case study, only the total combustion of methane (Reaction 38) was considered, where the increase of conversion goes from 34.53% in the basic design to 54.5% in the S3 scenario.

For the CZ, where the reforming reactions take place again, we can see an important variation with both Reactions 1 and 3 showing decreasing conversions, but additionally, we can observe that Reaction 1 has conversions greater than 100%, from 118.3% in the basic design, down to 97.72% in scenario S3; this behaviour is due to the reversibility of Reaction 3. As we can see in **Table 5**, its reaction equilibrium shifts to the left, which means that the CO_2 present in the mixture is consumed on average of −37% for all scenarios in conversion; therefore, additional methane is generated, and simultaneously it is also consumed by the Reaction 1 reaction, exceeding a 100% conversion; however, its output mass flow decreases as the substitution scenarios increase, which is due to (i) the increase of conversion in the PR, (ii) the reversibility of Reaction 3 in the CZ, and (iii) higher conversion of original methane and that formed in the SR. These conclusive points will be the basis for modifying the operating conditions of the process with the aim of reaching the final balance of CO_2.

4.2.2 HTS and LTS reactors

The WGS (2) reaction is the only one that takes place in the HTS and LTS reactors; the equilibrium of the reaction displaces to the right, forming CO_2 and steam with a high conversion percentage, for HTS 74.28% and LTS 91.6% on average for all scenarios. The conversion of Reaction 2 also increases as the percentage of replacement of natural gas by biogas grows.

The reversibility of Reaction 3 in the secondary reformer accelerates the conversion of Reaction 1, exceeding 100% due to the additional methane that is formed. This favours the formation of more CO that will enter HTS and LTS reactors and become CO_2 with highly efficient conversions. Therefore, the HTS and LTS reactors are also considered as control elements to reach the final balance of CO_2. Also note, that the CO flow production decreases as we raise the substitution scenarios.

4.3 Carbon dioxide CO_2 balance

Until now, we have described the kinetics behaviour of the reforming reactions as a function of the increase in CO_2 flow, due to the substitution with biogas in the feedstock, we must analyse how the total flow of this gas behaves from the feed point, its generation, up until the output of the syngas unit. **Figure 8** shows the simulated flowchart in Aspen Hysys V.14 with the set points defined to perform the CO_2 balance throughout the process unit.

Point 2 shows the total feed stream to the process including the mixture of natural gas and biogas in the different substitution scenarios. **Table 6** indicates the CO_2 mass flows defined according to **Figure 8**.

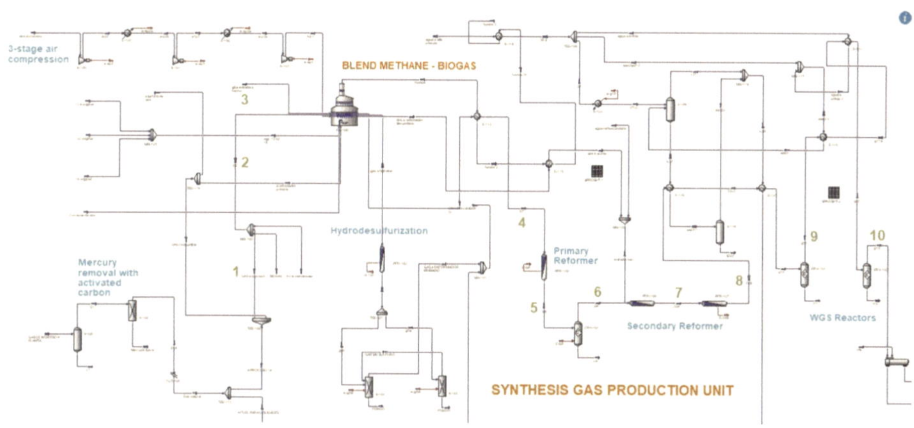

Figure 8.
Flow diagram of the syngas unit in Aspen Hysys V14.0. Where: (1) gas to process; (2) first feed to fire heater; (3) second feed to fire heater; (4) primary reformer inlet; (5) primary reformer outlet; (6) combustion zone, heavier hydrocarbons; (7) combustion zone, second reformer outlet; (8) catalytic zone, second reformer outlet; (9) HTS outlet; (10) LTS outlet.

Streaming yam	Basic design	S1 (70-30)	S2 (60-40)	S3 (50-50)
	lb/h	lb/h	lb/h	lb/h
(1) Gas to process	2928.48	2049.93	1757.09	1464.24
(2) First feed to fire heater	2928.48	13632.97	17201.13	20769.30
(3) 2nd feed to fire heater	3222.57	13961.67	17529.83	21098.00
(4) PR inlet	3222.57	13961.67	17529.83	21097.99
(5) PR outlet	36455.92	42591.27	44686.73	46804.32
(6) HZ, heavier hydrocarbons	56978.22	56956.86	57000.11	57065.47
(7) HZSR outlet	86921.55	86904.72	86949.44	87016.30
(8) CZSR outlet	63382.74	72676.62	75258.37	77530.06
(9) HTS outlet	122942.12	115981.20	113383.82	110364.44
(10) LTS outlet	145717.52	130070.69	124677.01	119222.85

Table 6.
Carbon dioxide CO_2 mass flow by scenario.

Stream name	Basic design	S1 (70–30)	S2 (60–40)	S3 (50–50)
	lb/h	lb/h	lb/h	lb/h
SYNGAS to absorption inlet	145437.33	129750.44	124274.34	118785.25
CO_2 absorption outlet	146097.71	130647.77	125154.10	119638.21
CO_2 to urea plant	145252.84	129758.42	124316.75	118854.62
CO_2 sub-product requirement	145252.84			
Difference	0	15494.42	20936.09	26398.22

The order of the colors is decreasing, where the darker shade indicates the higher result.

Table 7.
Flow difference at the outlet of the CO_2 removal unit.

First, we will analyse the behaviour of the CO_2 mass flow for the basic design and each one of the scenarios. In all cases, CO_2 increases up to the outlet in LTS, just at point (8, CZSR outlet), the CO_2 flow decreases because, in the catalytic zone of the secondary reformer, Reaction 3 is produced in the opposite direction, both consuming CO_2 and generating more amount of methane.

Now, if we compare the mass flow of CO_2 formed in the base design and the scenarios, we observe that the flow declines as the percentage substitution with biogas increases from 145717.52 lb/h in the base design to 119222.85 lb/h [10]. So, what is the reason for the decrease in CO_2 in the substitution scenarios? It would be expected that by increasing the supply of biogas, the net flux of CO_2 should be greater, but it is not.

The decrease in methane in the feed is due to the biogas composition, which is approximately 55% methane and 45% CO_2, and substituting natural gas that is 90% pure with the biogas composition. This influences a lower flow and production of CO_2 despite the fact that there is no major incidence in the conversion of the reaction as such. Later on, we will analyse the methane feed flow in the different scenarios in order to fully understand the decrease in CO_2.

Finally, the produced syngas must go through a series of absorption stages to remove the greatest possible amount of CO_2 that must be used as a secondary product. **Table 7** shows the missing difference of CO_2 in each study scenario compared to the basic design, based on the flow required at the outlet of the absorption process, so that it can be used as a secondary product (mass flow of CO_2 calculated in the basic design according to the nominal capacity of the plant); in the case of ammonia plants, CO_2 is used for the production of urea. The difference in flow becomes greater than the required specification (146,557 lb/h) as the percentage substitution of the scenarios increases, with a shortage of 26757.56 lb/h for the last S3 scenario. In Section 5, the operating conditions will be adjusted to compensate for the missing CO_2 flow, increasing the production of CO to use the WGS in the HTS and LTS reactors and produce a greater amount of CO_2.

4.4 Selectivity and molar equilibrium of the reforming reactions assessment

The final conversions of Reactions 1 and 3 in both reformers are mainly dependent on temperature. The growth in temperature increases the conversion in the reactors, particularly that of Reaction 1, but the molar equilibrium of the reaction interferes in the displacement from right to left, as in any equilibrium reaction. The global selectivity represents the produced amount of the desired product, in this case CO_2

	Basic	S1 (70–30)	S2 (60–40)	S3 (50–50)
PR	0.66	0.63	0.62	0.61
CZSR	0.53	0.52	0.54	0.59

Table 8.
Selectivity of CO_2 with respect to methane in syngas reformers.

	Reaction 1: $CH_4 + H_2O = CO + 3H_2$		Reaction 3: $CH_4 + 2H_2O = CO_2 + 4H_2$		
	Right side (lbmol/h)	Left side (lbmol/h)	Right side (lbmol/h)	Left side (lbmol/h)	
Base	11695.03	70.72	11695.03	143.95	PR*
	8655.97	6245.66	8655.97	7838.90	CZSR**
70-30	11162.09	70.74	11162.09	387.97	PR
	8711.51	5349.49	8711.51	6939.71	CZSR
60-40	10985.13	70.76	10985.13	469.05	PR
	8873.50	5047.66	8731.70	6638.90	CZSR
50-50	10808.17	70.77	10808.17	550.13	PR
	8753.39	4738.41	8753.39	6336.12	CZSR

*PR = primary reformer. **CZSR = catalytic zone, second reformer.*

Table 9.
Molar equilibrium of the Reactions 1 and 3 in the PR and the CZSR.

regarding the reacted amount of methane. **Table 8** shows the CO_2 selectivity in the primary reformer and the catalytic zone of the secondary reformer.

The global selectivity of CO_2 considers that Reactions 1 and 3 occur in syngas reformers. The basic design presents a selectivity of 66% in PR and 53% in the CZSR, and the selectivity of CO_2 decreases as the percentage of substitution increases, although it does not represent a significant decrease.

The molar equilibrium of the reforming reactions is observed in **Table 9**, this is oriented both in all scenarios and both reactors to the right, which means that the formation of CO_2 is permanent in all cases, but when we analyse the conversion percentages, we note that in the CZSR, reversibility of Reaction 3 occurs, and methane is formed; additionally the molar flows on the left and right do not exhibit a very large molar difference compared to RP, therefore, we can conclude that the total conversion in the ZCSR is mainly controlled by the temperature of the reactor. Finally, we will analyse the differential flow of methane that is in the feed to the process because the composition of biogas with 61% is less than that of the content in natural gas with 91%.

As previously described, the composition of methane in the feed to the process is 92% for natural gas and 61% for biogas. This variation affects the amount of total methane that enters the syngas reactors, which is lower as the substitution scenario increases. For the basic design, the flow of methane feed is 50114.6 lb/h, and for the study scenarios is as follows:

- S1. Total methane mass flow feed: 41597.6 lb/h
- S2. Total methane mass flow feed: 38758.6 lb/h
- S3. Total methane mass flow feed: 35919.6 lb/h

The decrease in methane available to react in each scenario does not affect the conversion of chemical reactions, but it does have a negative effect on the CO_2 balance, since the flow decreases with the increase in percentage replaced. For example, the shortage in S3 is 26757.56 lb/h. This amount of missing CO_2 leads us to conclude that in S3, even modifying the operating conditions by applying severity factors, it will not be possible to reach the required flow specification as a secondary product; therefore, S3 will not be considered for the subsequent study of the modification of operational variables.

5. Modification of operational variables

Simulations were developed with modification of variables in scenarios S1 and S2 with the objective of increasing the conversion of Reaction 1 to obtain more CO and, in consequence, greater flow of CO_2 in the HTS and LTS reactors. Considering the following:

- The first change of variable focused on the severity of the temperature, increasing to 1400°F, 1454°F and 1508°F, respectively, at the exit of the primary reformer and maintaining the original exit temperature of the secondary reformer.
- For the second part, it was considered to simulate the adiabatic secondary reformers; therefore, the secondary reformer outlet temperature is determined by the severity of the primary reformer and the amount of air entering the process.

5.1 Scenario S1: 70-30

Table 10 shows the results obtained with the increase in temperature at the outlet of the primary reformer without exceeding the severity parameters of the equipment.

Reactor	Temperature	1400°F	1454°F	1508°F
PR	Reaction 1	23.44	30.00	37.14
	Reaction 3	26.84	27.37	27.42
HZSR	Reaction 38	52.59	61.35	73.76
CZSR	Reaction 1	118	130.2	158.1
	Reaction 3	−59	−89.99	−164.1

The order of the colors is decreasing, where the darker shade indicates the higher result.

Table 10.
Temperature increase in the primary reformer for S1.

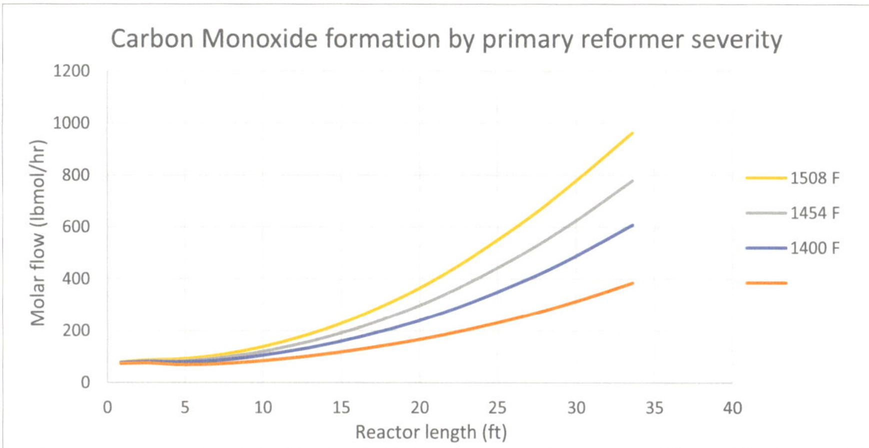

Figure 9.
Carbon monoxide formation by primary reformer severity.

Reaction 1 shows conversion increase with respect to the original S1. At 1400°F the conversion increase is 58.06% higher, 102.9% at 1454°F and 150.44% at 1508°F.

Reaction 3 also presents an increase in conversion, although the proportion is lower compared to that of Reaction 1, reaching a maximum increase of 2.33% at a temperature of 1508°F, so we can conclude that the increase in temperature in the primary reformer has a greater incidence in Reaction 1. The case study is a positive factor since we seek to increase the production of the WGS reaction in the HTS and LTS reactors. **Figure 9** shows the increase in CO production with the increase in temperature compared to S1.

In original S1, the produced CO flux is 10768.47 lb/h at the outlet of the primary reformer. For the temperature of 1400°F, the flow of CO increases by 6258.37 lb/h, 11,024 .34 lb/h for 1454°F and 16206.39 for 1508°F. With an increase in temperature in the primary reformer, higher conversions of Reaction 1 are reached, which significantly favours the production of CO.

If we look at the consequences of the first change on the secondary reformer, the Reaction 38 conversion increases with temperature, reaching a maximum value of 73.76% at a temperature of 1508°F, although from the CO_2 production approach, high conversion is favourable, the methane remainder will have an impact on the reactions of the ZCSR, this presents the same behaviour as the original S1, but the conversions are large, considering that the remaining methane flow that enters to react is 9804.4 lb/h; clearly these results indicate that the conversion does not necessarily reflect a high production, on the contrary, the increase in temperature promotes the formation of coke.

For example, at a temperature of 1508°F, Reaction 1 reaches 158% and −164.1% in reversibility to Reaction 3 for a methane inflow of 3868.43 lb/h, considering that the remaining methane at the CZSR outlet is adjusted to 1.72%, then a representative contribution in the conversion change is not observed. Therefore, in order to consider the second modification variable (SR adiabatic condition), only the temperatures of 1400°F and 1454°F will be taken into account.

Table 11 shows the results of the modification of the two conditions: (i) increase in the severity of the primary reformer, and (ii) adiabatic secondary reformer; at the temperatures selected based on the above analysis.

Reactor	T (°F)	1400	1454
HZSR	Outlet T. (°F)	2071.4	2156
CZSR (adiabatic)	Reaction 1 ZC	127.7	132.6
	Reaction 3 ZC	−28.58	−33.25
	CH_4 mole fraction outlet	0.002	0.0001
	CO outlet (lb/h)	38911.45	37681.56

The order of the colors is decreasing, where the darker shade indicates the higher result.

Table 11.
Adiabatic secondary reformer for S1.

The conversion of Reaction 1 increases by 9.7% with respect to the result of **Table 11** up to 127.7%. On the contrary, the inverse conversion of Reaction 3 decreases down to −28.58%, which means that less accumulated CO_2 is consumed. One of the most important effects of the adiabatic change of the secondary reformer is the almost total consumption of methane. The mole fraction at the CZSR outlet is 0.0002 for 1400°F and 0.0001 for 1454°F, with the average outlet flow being 82.82 lb/h.

The final flow of CO obtained at 1400°F is 38911.47 lb/h, and at 1454°F 37681.56 lb/h enters the HTS reactor. With these results, the conversion of the Reaction 2 reaction remains the same, but having a greater flow of CO fed, the production of CO_2 increases in 10522.05 lb/h more than the original S1, up to 140642.16 lb/h. For the temperature of 1454°F the conversions in CZSR are higher; however, the final flow of CO_2 obtained at the exit of the LTS is 140742.32 lb/h, only 100 lb/h more than the result obtained at 1400°F. In conclusion, temperatures above 1400°F are not justified to achieve higher CO_2 production.

5.2 Scenario S2: 60-40

We apply the same procedure used in S1, considering the modification of the variables in two steps. **Table 12** shows the results of the temperature increase for S2.

The behaviour is the same as in S1. Reaction 1 increases its conversion to a maximum of 38.79% at the temperature of 1508°F. For Reaction 3, no relevant increase in conversion is observed with increasing temperature, and the conversion reaches its equilibrium limit at 27.8%. The CO flux at the RP outlet increases by 6080.58 lb/h,

Reformer	T (°F)	1400	1454	1508
PR	Reaction 1	24.82	31.58	38.79
	Reaction 3	27.25	27.73	27.78
HZSR	Reaction 30	58.54	68.97	83.95
CZSR	Reaction 1 ZC	113.5	127.3	173.8
	Reaction 3 ZC	−64.05	−106.7	−260.7
	$T_{S\,ZC}$ (°F)	1565.6 F	1532.3 F	1495.4 F

The order of the colors is decreasing, where the darker shade indicates the higher result.

Table 12.
Temperature increases in the primary reformer for S2.

Reactor	**T (°F)**	**1400 F**
HZSR	Outlet T. (°F)	2149
CZSR (adiabatic)	Reaction 1 ZC	116.4
	Reaction 3 ZC	−16.88
	CH_4 mole fraction outlet	0.0001
	CO outlet (lb/h)	32449.94

Table 13.
Adiabatic secondary reformer for S2.

Stream name	**Basic design**	**S1 (70-30) T = 1400°F**	**S1 (70-30) T = 1454°F**	**S2 (60-40) T = 1400°F**
	lb/h	**lb/h**	**lb/h**	**lb/h**
(11) Syngas to absorption inlet	145437.33	140295.42	140397.31	134537.52
(12) CO_2 absorption outlet	146097.71	141244.37	141323.27	135484.40
(13) CO_2 to urea plant	**145252.84**	**140254.24**	**140334.47**	**134559.33**
CO_2 sub product requirements	**145252.84**			
Difference	0	4998.60	4918.37	10693.51

*Improved scenarios represent increased severity in PR and change to an adiabatic SR.**
The order of the colors is decreasing, where the darker shade indicates the higher result.

Table 14.
Flow difference at the outlet of the CO_2 removal unit with improved scenarios.

10652.33 lb/h, and 15532.84 lb/h, respectively, for each temperature, compared to the original S2.

The conversion of Reaction 38 in HZSR increases with the temperature up to a maximum of 83.95% at a temperature of 1508°F; as we know, the CZSR presents the same behaviour and the remaining methane that enters is 7698.08 lb/h for 1400°F, 4893.49 lb/h for 1454°F and 2078.18 lb/h for 1508°F; therefore, no net syngas production is observed differing significantly from the original S2. With the increase in severity, only the risk of coke production is valued without a significant benefit in the production of syngas.

To consider the second modification variable (SR adiabatic condition) in S2, only the temperature of 1400°F is considered.

The conversion of Reaction 1 increases 3% with respect to the result of **Table 13**, up to 116.4%. On the contrary, the inverse conversion of Reaction 3 decreases by 47.17%. The high decrease in the inverse conversion of Reaction 3 indicates that the formation of methane drops considerably in the CZSR. For the operation of **Table 14** in the configuration of the adiabatic secondary reformer, it was observed that the methane consumption limit is 0.0001 in mole fraction; thus, the methane conversion reaches 99%.

With these results, the flow of CO formed is 32449.94 lb/h. If we compare it with the flow of CO in the original S2, the difference is 402.78 lb/h more. At the exit of LTS, carbon dioxide production is 134916.80 lb/h, that is, the production increase is 10239.77 lb/h, compared to the original S2.

6. Conclusions

From the scenarios where natural gas is replaced with biogas, we can confirm that the presence of greater amounts of CO_2 in the feedstock to the processing unit directly affects the increase in the conversion of Reaction 3 in the reformers; however, by having a lower flow of methane fed, the total net produced CO_2 balance decreases as the replacement percentage increases.

The formation of methane is undoubtedly a key element for the selectivity towards CO to be favoured in Reaction 1 due to the reversibility of Reaction 3 in the CZSR.

From the results obtained, we can say that the S3 scenario declined for the adjustment of operational variables because despite the increase in severity and the adiabatic SR, the production of enhanced CO would not be sufficient to reach the CO_2 demand required to produce urea.

The S1 and S2 scenarios were the basis for the change of operating variables to increase the production of CO_2 in the syngas unit. An increase in temperature in PR and the change to an adiabatic SR, directly promote the formation of CO (Reaction 1), therefore, increasing the production of CO_2 in the HTS and LTS reactors. In **Table 15**, we can see how the CO_2 deficit decreases, and therefore, the gap that was found in **Table 14** for each substitution scenario.

So, we can see that for S2 at 1400°F, the CO_2 deficit decreases by 15,705 lb/h, and although the production is positive, the missing flux is considerable; therefore, it is not considered a favourable scenario for the technical and potential economic requirements of the plant.

In the case of S1, at both temperatures, the CO_2 deficit does not exceed 5000 lb/h reaching the required net balance is feasible only by slightly decreasing the flow of biogas in the feed, and since there is no significant variation between the flow of CO_2 produced at 1400°F vs. 1454°F, the most recommended scenario is S1 at 1400°F. It should not be forgotten that the increase in temperature has an impact on an increase in fuel gas required for the plant.

It is important to remember that the study considers the substitution of fossil natural gas, only for the feed to the syngas process (not the natural gas required as fuel). Therefore, the CO_2 equivalent balance will show that, with the adjustments

CO_2 equivalent balance				
Scenario	**IPCC 2021 (lb/h)**			
	Total, from inlets	**Total, from outlets**	**(13) CO_2 to urea plant**	**CO_2 eq Scope 1**
Basic design	2358085.76	793208.61	145252.84	−1710129.99
S1 (70-30)	2130325.44	777626.33	129758.42	−1482457.54
S1 (70-30) T = 1400°F	2130325.44	786944.11	140254.24	−1483635.58
S1 (70-30) T = 1454°F	2130325.44	787014.45	140334.47	−1483645.46
S2 (60-40)	2054393.80	772198.81	124316.75	−1406511.74
S2 (60-40) T = 1400°F	2054393.80	781233.42	134559.33	−1407719.71

Table 15.
Balance of CO_2 eq for original scenarios and with modification of variables.

made to the process, the most significant contribution to GHG emissions is due to the unit's fuel.

The CO_2 eq balance considers the input and output flows to the processing unit to calculate the net value of emissions in scope 1. **Table 15** shows the balance made based on the original scenarios and the variation of the selected operating conditions. The CO_2 eq balance was calculated in Aspen Hysys V.14 according to the IPCC 2021.

It is important to mention that for the balance developed, the flow of CO_2 eq from the biogas (CO_2 capture process) was not discounted from the feed. By considering his case in the balance, the input flows would be even lower, mainly due to fuel gas feeds to the process.

The total input flow of CO_2 eq decreases as the percentage of substitution with biogas increases. This is mainly due to the decrease in the flow of fossil natural gas that, as we know, has components with a much higher GPW.

All the output flows in Scope 1 are lower than those of the original design, and we also observe that the modification of the operating variables slightly increases the output flow for each scenario. This result is to be expected due to the increase in CO_2 production.

Finally, the balance of Scope 1 is negative for all cases, and this means that CO_2 is captured. For example, for an ideal case (S1 at 1400°F), we have 1483635.58 lb/h of CO_2 eq avoided. All the scenarios, both studied and modified, show good negative figures, but in considering a balance between the best Scope 1 achieved and the efficiency of the process itself, such as the case of S2 at 1400° F, which has a lower flow of CO_2 avoided (for the decrease in fossil fuel), the CO_2 deficit in the process is very high and cannot be covered. In addition to this, the required fuel and the probability of forming coke with the increase in temperature must be considered. Therefore, S1 at 1400°F is the most favourable outcome in this study.

List of acronyms and abbreviations

ATR	autothermal reforming
BG	biogas
CZSR	catalytic zone secondary reformer
GHG	greenhouse gas emissions
GWP	global warming potential
HTS	high temperature shift reactor
HZSR	high combustion zone secondary reformer
IPCC	Intergovernmental Panel on Climate Change
NG	natural gas
LCA	life cycle assessment
LTS	low temperature shift reactor
SMR	steam methane reforming
POX	partial oxidation
Syngas	synthesis gas
PR	primary reformer
SR	secondary reformer
TRLs	Technology readiness levels
TS	outlet temperature
WGS	water gas shift

Author details

Martha Siles Camacho* and Valeria Sandoval Cossío
Universidad Privada del Valle, Cochabamba, Bolivia

*Address all correspondence to: msilesc@univalle.edu

References

[1] United States Environmental Protection Agency. EPA [Online]. 2023. Disponible en: https://www.epa.gov/ghgemissions/understanding-global-warming-potentials

[2] Sandoval Cossio VE. Modelo de estudio del proceso cinético de reformación de gas natural a syngas basado en distintos escenarios operativos para la producción de amoniaco con tecnología Kellog Brown & Root en Bolivia. Cochabamba, 2023

[3] Oficina Española de Patentes y Marcas. Patente Europea 2 547 428 [Online]. 2015. Disponible en: https://patentimages.storage.googleapis.com/2f/9b/4e/0b76cf9b88daf0/ES2547428T3.pdf

[4] Moulijn JA, Van Diepen AE. Chemical Process Technology. Germany: John Wiley & Sons, Inc.; 2013

[5] Stafford W, Lotter A, Brent A, von Maltitz G. Biofuels Technology. The United Nations University World Institute for Development Economics Research (UNU-WIDER); 2017

[6] National Academies of Sciences, Engineering, and Medicine. Negative Emissions Technologies and Reliable Sequestration. The National Academies Press; 2019

[7] Speight JG. Handbook of Petrochemical Processes. Estados Unidos: CRC Press; 2019

[8] Speight J. Handbook of Industrial Hydrocarbon Processes. 2nd ed. Cambridge, MA: Gulf Professional Publishing is an imprint of Elsevier; 2020

[9] Osejos Merino M, Jaramillo Véliz J, Merino M, Quimis Gómez A, Alcívar CJ. Producción de biogás con estiércol de cerdo a partir de un biodigestor en la. Dominio de las ciencias. 2018;**4**(1):709-733

[10] Salazar Del Pozo Luis Fernando. Estudio comparativo de métodos de remoción de dióxido de carbono y sulfuro de hidrógeno en el biogas producido en la unidad experimental de producción de biogas de univalle para potenciar sus aplicaciones. Cochabamba; 2022

[11] Piña. Simulación y Optimización de Reactores de Reformado de Metano con Vapor. Obtenido de Tesis Doctoral en Ingeniería Química. 2003. Available from: https://repositoriodigital.uns.edu.ar/bitstream/handle/123456789/2187/Pi%C3%B1aJuliana-Tesis.pdf?sequence=1&isAllowed=y

[12] Aasberg-Petersen K, Dybkjær I, Ovesen CV, Schjødt NC, Sehested J, Thomsen SG. Natural gas to synthesis gas – Catalysts and catalytic processes. Journal of Natural Gas Science and Engineering. 2011;**3**(2):423-459. DOI: 10.1016/j.jngse.2011.03.004

[13] Nielsen R, Christiansen L. Concepts in Syngas Manufacture. London: Imperial College Press; 2011

[14] Xu J, Froment GF. Methane steam reforming, methanation and water-gas shift: I. Intrinsic kinetics. AIChE Journal. 1989;**35**(1):88-96. DOI: 10.1002/aic.690350109

[15] Liu K, Song C, Subramani V. Hydrogen and Syngas Production and Purification Technologies. New Jersey, Hoboken: John Wiley & Sons, Inc.; 2010

Chapter 4

Perspective Chapter: Determinants of Natural Gas Prices in the United States – A Structural VAR Approach

Dixon Domfeh

Abstract

This study examines the dynamic economic relationships between the fundamental variables that influence natural gas prices within the U.S. market. A structural vector auto-regressive (VAR) and Markov switching models were used to investigate the impact and stability of regime switches between the main drivers of natural gas prices. The result reveals that the U.S. gas market market is sensitive to temperature deviations in the short term. Crude oil and coal prices have long-run effects on natural gas prices, emphasizing energy-specific demand shocks. Mainly, coal prices determine about 73% of gas price variability for the sample period with three discernible regime switches.

Keywords: natural gas, structural VAR, Markov switching, weather, storage

1. Introduction

This study explores the dynamic interplay among the core determinants of natural gas pricing. The cost of natural gas holds significance for multiple stakeholders given its pivotal role in the heating and cooling industry, as well as its use in the generation of electricity. Consequently, understanding the natural gas price formation is crucial from both macroeconomic and micro-economic viewpoints. Nonetheless, the evolution of pricing is intricate due to numerous distinctive supply and demand traits that markets confront, including weather conditions, the effects of cross-commodity substitution, unexpected changes in storage, and fluctuations in business cycles.

It is difficult to disentangle the spillover effect of energy commodities due to their interconnected usage. For instance, natural gas could be used as an energy input to substitute crude oil in electricity production and heating and cooling in residential quarters. Xu Gong and Wang [1] have analyzed the volatility spillover effects among the four major commodities (i.e., crude oil, gasoline, heating oil and natural gas) in the U.S. futures market and found that spillovers have peaks and troughs during pivotal periods in history such as the U.S. shale gas discovery, the Great Financial Crisis and the oil price crash. Wang et al. [2] show that increasingly financial market factors such as investor speculation are becoming important in natural gas price formation. Notwithstanding the financialization of energy commodities, there are other fundamental factors that affect the price formation of commodities. More specifically, natural gas may have regional and international factors that drive price formation. Zhang and Ji [3] and Wang et al. [4] emphasize that natural gas markets have strong regional characteristics. As such research findings from different regions of the world such as Europe, China and the U.S. may not always be consistent. For example, Nick and Thoenes [5] find that storage conditions have a significant effect on natural gas prices following a logical economic narrative in Germany. Elsewhere, Wang et al. [2] find that storage conditions have marginal input in natural gas price movements in the U.S. The regional discrepancies may arise for several reasons. In the U.S. electricity providers who use most of the natural gas have operation mechanisms that allow them to switch between inputs (i.e., natural gas and coal). The regional price formation of natural gas prices can also be attributed to liberalization of hubs. Wang et al. [4] show that the Chinese market's price of natural gas trades at a premium because of its link to oil indexation and not necessarily driven by market fundamentals. Specifically investigating the volatility in prices, Hailemariam and Smyth [6] report that gas prices are responsive to structural supply shocks in the U.S. citing several prominent events such as Hurricane Katrina, Global financial crisis and the discovery of shale resources. Moreover, technological and climate change polices continue to evolve the landscape of the gas market in the U.S.

In this research, five fundamental drivers of natural gas prices are identified in the U.S. gas market. The supply and demand interactions of temperature and storage deviations, short-term interest rate, coal, and crude oil prices are analyzed as variables that affect natural gas prices. We develop a structural vector autoregressive model (VAR) and a Markov Switching process to investigate the phenomena. The models yield insights into the dynamic relationship between the variables under study. The impulse response simulations from the VAR model are consistent with economic reasoning, implying that natural gas prices react to underlying supply and demand shocks. Natural gas prices rise in reaction to abnormal temperatures, which increases heating and cooling demands in the short term. The response of gas prices to structural shocks of storage is insignificant from this study. However, the empirical results reveal that storage level depletes through withdrawals in response to extraordinary temperatures, which increases demand for heating or cooling. Coal and crude oil price shocks show long-term effects on the natural gas price formation.

A significant contribution of this research is identifying the distinct contributive effects of the different variables on natural gas prices. It is interesting to find coal prices determining about 73% of gas price variability for the sample period, unambiguously making it a systematic determinant of natural gas prices. A three-state Markov switching process between coal and gas prices has been identified with noticeable regime switches.

The discovery that the cost of coal is the main determinant of natural gas prices could question the conventional notion that crude oil is the key factor in determining gas prices. For example, Hartley and Medlock [7] and Brown and Yucel [8] argue that gas and crude oil prices are co-integrated in the short and long run. However, the stability of this cointegration dynamics is not stable over time. Wang et al. [2] and Ramberg and Parsons [9] bring to bear that the cointegration relationship between oil and gas in the U.S. market decouples over time. Particularly, Wang et al. [2] demonstrate the declining role of oil prices in natural gas pricing mechanism with time-series data between 2001 and 2018. This is due to the demise of crude oil indexation of natural gas prices. The findings of this paper is consistent with an earlier work by Nick and Thoenes [5] who find a similar strong evidence of connection between coal and natural gas prices in Germany.

In the past, the cost of crude oil has been perceived as the most influential element impacting long-term natural gas prices. A straightforward guideline, known as the 10-to-1 rule, was dominant in the 1990s. This rule proposed that the price of natural gas was one-tenth that of crude oil, as indicated by Brown and Yucel [8]. This trend of linking prices to oil has shifted in recent years, particularly in the United States, as a result of the shale gas revolution that started in 2005. With the successful application of horizontal drilling and fracking technologies, shale gas production has increased exponentially, accounting for about 79% of total gas supplies in the U.S.

The description of the data used are discussed in Section two. Section three outlines the empirical methodologies employed in previous research on the natural gas market, definition, and identification of the structural VAR model. Empirical results in the form of impulse responses and Markov switching processes are explained in Section four. A summary and conclusion are presented in Section five.

2. Data description

The dataset used in this research consists of 471 observations of weekly data within the period from January 2, 2010, to December 29, 2018.[1] It comprises the Henry Hub natural gas spot price, West Texas Intermediate (WTI) crude oil spot price, average U.S. spot coal price, U.S. natural gas storage, the annual yield on 3-month U.S. Treasury Bill, and the consumption-weighted temperature deviations from historical averages for the top 5 natural gas consumption states in the U.S.[2] **Figure 1** displays all the time series of the variables and **Table 1** provides a summary definition of the variables used in this study.

The choice of variables for this econometric analysis is comprehensive as it encompasses the diverse fundamental drivers of natural gas prices both on the demand and supply side. The variables used in this paper have been well studied and documented in the energy economics literature (see for examples [2, 10–13]). Variables such as exchange rates are not applicable to the U.S. market it is a net exporter of natural gas. We do not consider the role of speculative investors to be a fundamental factor in this paper since we feel that they are more related to the financialization aspect of commodity market as a whole and not specific to natural gas. Spot prices are heavily relied upon due to their ability to capture significant short-term impacts, particularly those of interest, such as demand spikes triggered by temperature variations. Furthermore, Zhang and Ji [3] suggest that for the U.S. market hub prices can reflect the true value of natural gas. Here, the Henry Hub spot price was used because its price movement is considerably affected by the temperature conditions of the top 5 natural gas consumption states considered in this study.

The model was calibrated using weekly data to allow for the inclusion of coal price data, which is only available in weekly frequency while capturing short-term meteorological conditions. Prices of crude oil and coal are used as proxies for energy-specific demand. The inclusion of the spot price of crude oil is essential for capturing the interdependence between oil and gas in the domestic heating and cooling market. By

---

1 The observations are averages for each week ending on Friday.

2 The top 5 natural gas consumption states are Pennsylvania, Florida, Texas, Louisiana and California. The weights are as follows: California, 0.20; Florida, 0.13; Louisiana, 0.15; Pennsylvania, 0.12; Texas, 0.37.

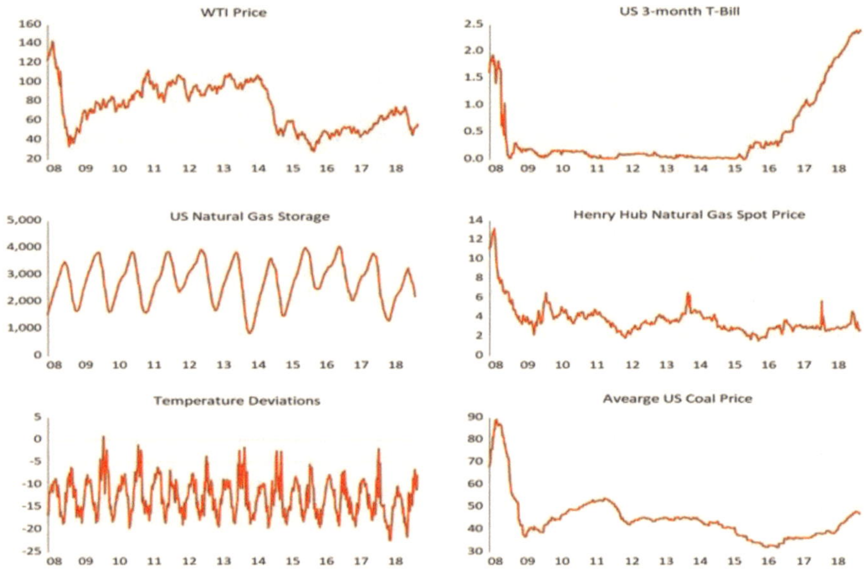

Figure 1.
Time series of all variables used.

incorporating coal prices into the model, the interaction between gas and coal in the electricity generation sector can be captured. This enables the representation of cross-commodity effects associated with fuel substitution (see, for example, [5]). Natural gas, coal, and crude oil price time series were transformed to their natural logarithms. This approach is consistent with macroeconomic literature (see, for example, [14]). The

Variable	Description	Unit	Source
Temperature deviations	Measured as the degrees day (D.D.) deviations from historical averages. DD = CDD + HDD. Where CDD is cooling degrees day and HDD is heating degrees day	Degrees Fahrenheit	National Weather Service (NWS)
WTI price	West Texas Intermediate spot price for U.S. crude oil	USD	
Per barrel	Bloomberg		
Coal price	The average spot price of U.S. coal from the five producing regions (Central Appalachian, N. Appalachian, Illinois Basin, Powder River Basin, Uinta Basin)	USD per ton	Quandl
Natural gas price	Henry Hub spot price	USD per million Btu (British thermal unit)	Bloomberg
Storage	Working gas in underground storage	Bcf (Billion cubic feet)	U.S. Energy Information Administration
Treasury bill	Three months U.S. treasury bill	Percent	Federal Reserve Bank of St. Louis

Table 1.
Summary description of variables used.

structural VAR (vector autoregressive) equation was estimated with log-level prices because the major interest is in the dynamic economic relationships between the selected variables with the natural gas market and not any possible cointegration or stationarity.

The natural gas demand in the residential heating market is susceptible to temperature. In a liberalized market, it is anticipated that storage operators will take advantage of predictable seasonal demand fluctuations. Hence, exceptional short-term weather conditions that result in a shift in demand are anticipated to be significant factors in gas price formation, rather than predictable seasonal patterns. Consequently, the degree day deviations (D.D.) from normal historical averages was equally constructed to determine gas prices. Temperature expressed in degree days is a quantitative index that reflects a demand for energy to heat or cool houses and businesses. Degree days (D.D.) is the sum of heating degree days (HDD) in the winter and cooling degree days (CDD) in the summer.[3]

$$DD_t = CDD_t + HDD_t \tag{1}$$

$$CDD_t = Max(0, Tave_t - 65^o F) \tag{2}$$

$$HDD_t = Max(0, 65^o F - Tave_t) \tag{3}$$

where $Tave_t$ is the average daily temperature of date t. The weather shock is defined as the deviation of D.D.s from the normal level over the period.

$$W_t = \sum_{i=1}^{m} (DD_{t+i} - DDNORM_{t+i}) \tag{4}$$

where m is the weather period, DD_{t+i} is the degree days on day $t+i$, $DDNORM_{t+i}$ is the normal degree days which is the average degree days of the previous 30 years on day $t+i$. The use of deviations of the observed D.D. from their historical averages estimates the effect of unexpected weather conditions on gas prices.

In this study, storage data was included because storage operators are both on the supply-side (withdrawal phase) and demand-side (injection phase). Furthermore, instead of absolute volumes, the change in utilization rate was employed as an indicator of fluctuations in total storage capacity. The discrepancy between the average seasonal changes in utilization and the actual weekly changes was used as a proxy for the capacity of flexible storage to respond to storage shocks. The deep reasoning for this approach is to capture deviations from the seasonal storage utilization pattern. Natural gas production was not included since the U.S. has been a net exporter of LNG and crude oil since the shale gas boom in 2006. Three months treasury bills was fitted as a control variable. Treasury bills may affect natural gas prices because the short-term interest rate is a significant component of the cost of carrying inventories [15, 16].

3. Empirical methodology

There has been enormous empirical research on the natural gas market in recent years. A common thematic approach underlying these findings is the relationship between natural gas and crude oil prices rather than exploring how the natural gas

[3] HDD and CDD are widely used weather derivatives in the energy industry which are traded at the Chicago Mercantile Exchange (CME).

market works. These approaches ignore the possibility of structural dynamics within the market. For example, [7, 8, 17–19] find movements in oil prices to influence natural gas prices. Brown and Yucel [8] analysis involved testing for a cointegrating relationship using a vector error correction model (VECM). In contrast, other studies conclude that there is a weak link in the so-called oil-gas relationship (see, for example [3, 9]).

Other complex and comprehensive frameworks have been used to explore and disentangle the supply and demand shocks driving natural gas prices using various structural or reduced-form models. Nick and Thoenes [5] analyze the German gas market using a structural VAR and find that weather, supply, storage, coal, and oil price shocks have a significant effect on natural gas price formation. Hou and Nguyen [20] employ a Markov switching structural VAR model to investigate the regime-dependent responses to its fundamental shocks. They find that the impact of oil on natural gas prices is relatively small and regime-dependent. Our attempt to use a structural VAR to model the fundamental drivers of weather, supply, interest rate, crude oil, and coal prices is a novel approach in U.S. gas market research.

3.1 Model definition

We utilized a structural vector autoregression (SVAR) methodology to model the interrelationships among key factors in the natural gas market. This approach allowed us to explicitly analyze the transmission channels that influence natural gas prices. To constrain the feedback effects of exogeneity of some of the variables used in this analysis, their coefficients was restricted to zero. The VAR model in its reduced form can be represented as:

$$y_t = \upsilon + A_1 y_{t-1} + \ldots + A_p y_{t-p} + \varepsilon_t, \varepsilon_t \sim WN(0, \Sigma) \tag{5}$$

where $y_t = \left(y_t, \ldots \ldots, y_{m,t}\right)'$ is a vector of m endogenous variables, p is the number of lags of each variable, and υ is an $m \times 1$ vector of intercepts.

The error terms are grouped into $\varepsilon_t = (\varepsilon_t, \ldots, \varepsilon_{m,t})'$, such that the error term in each equation has a zero mean and is uncorrelated over time and homoskedastic. However, the error terms can be contemporaneously correlated with errors in other equations. Therefore, ε_t is a multivariate white noise process, $\varepsilon_t \sim W.N.(0, \Sigma)$, where Σ is an $m \times m$ variance-covariance matrix. The VAR model was specified with a length of six lags as identified by the Akaike Information Criterion. The SIC indicates a lag length of one; however, there is a strong autocorrelation in the error terms with one lag.

The instantaneous causality among the variables represented by ε_t in the reduced-form model does not allow for an economic interpretation of the error term. To remedy the situation, a structural model has to be identified as:

$$y_t = (I_m - A)y_t + A^* y_{t-1} + \ldots + A^* y_{t-p} + \varepsilon_t \tag{6}$$

where I_m represents the identity matrix of order m, A is an $m \times m$ matrix of instantaneous interaction among the variables and A^* is equal to $AA_{(i)}$ for $i = 0, \ldots, p$. Additionally, $\varepsilon_t = (\varepsilon_{1t}, \ldots, \varepsilon_{m,t})'$ is a row vector of m dimension which represents the structural errors with a variance-covariance matrix Σ_ε. As the instantaneous causality is captured by A, Σ_ε is diagonal. By assigning the errors of the structural representation to a specific variable, it becomes possible to interpret them in accordance with economic theory. It enables a deeper understanding of the underlying economic relationships and their impact on the variable of interest.

3.2 Model identification

To identify the model, restrictions are placed on the instantaneous coefficient matrix A. The structural representation can be achieved by imposing $m(m+1)/2$ restrictions. The instantaneous restrictions imposed for identification of the structural VAR is summarized in **Table 2**.

Since weather and interest rate are exogenous to the other variables in this analysis, they are ordered first within the matrix of contemporaneous interactions. Temperature deviations are expected to affect storage levels (via withdrawal from underground working gas) for cooling or heating purposes. Consequently, the instantaneous influence of temperature on storage unrestricted was ignored.

Storage of a commodity like natural gas is expected to be affected by the prevailing interest rate due to the cost of carrying inventory. In a high-interest rate environment, storage cost increases; therefore, the interaction between interest rate and storage was left intact. Gas price changes are expected to influence gas storage as storage operators primarily engage in inter-temporal price arbitrage. This economic rationale may drive the decisions and behavior of storage operators, emphasizing the significant impact of gas price fluctuations on storage dynamics. Furthermore, gas storage serves as a means to address temporary imbalances in supply caused by unexpected market conditions, including extreme weather events. It provides a mechanism to mitigate the effects of such disruptions and maintain stability in the gas market. Thus, contemporaneous impact of oil, coal, natural gas prices on storage is left unrestricted.

Enabling instantaneous interaction between crude oil, coal, gas prices, and temperature deviations is a logical approach. Extreme cold temperatures increase the demand for heating oil and can raise the price of WTI through this channel. The price of coal may be affected by oil and gas prices since coal serves as a substitute for electricity production. Therefore, the contemporaneous interaction between coal, gas, oil prices, and temperature are unrestricted.

The effect of interest rate across all three commodity prices (i.e., crude oil, coal, and natural gas) is unrestricted as it may have a significant explanatory power on the price variability. The overshooting model of Dornbusch hypothesizes that monetary changes have real short-run effects on commodity prices [21]. Whenever markets

	Temperature	Interest rate	Storage	Crude price	Coal price	Gas price
Degree days deviations	*	0	0	0	0	0
Change in 3-month T-bill	0	*	0	0	0	0
Storage deviation	*	*	*	*	*	*
Price of WTI	*	*	0	*	0	0
Price of coal	*	*	0	*	*	*
Natural gas price	*	*	*	*	*	*

*Notes: Each row represents an equation in the VAR model with the independent variables ordered in columns. The * denotes the parameters estimated from the data and allow for instantaneous interaction between the variables, whereas 0 indicates that the parameter has been restricted to zero.*

Table 2.
Identification of the contemporaneous matrix.

adjust in response to monetary changes, those sectors of the economy that are free to move (i.e., commodities and financial assets) must bear the burden of the sluggishness of the sticky sectors (i.e., manufacturing and services).

In this research, the focus is primarily on the price of natural gas as the main variable of interest. Therefore, no restrictions are imposed on that variable in the equation. This approach allows for a comprehensive analysis of the immediate impacts of all variables included in the model on natural gas prices. By considering all variables, we can understand the factors influencing natural gas prices.

4. Empirical results

4.1 Impulse response function

The application of the VAR model concerns the evaluation of the impact of structural shocks, which can be done using the moving average (M.A.) representation. **Figure 2** presents the estimated impulse response functions for the natural gas price.

The response of gas prices to changes in temperature, as observed in the impulse analysis, aligns with economic narrative. Extreme temperature has an immediate and

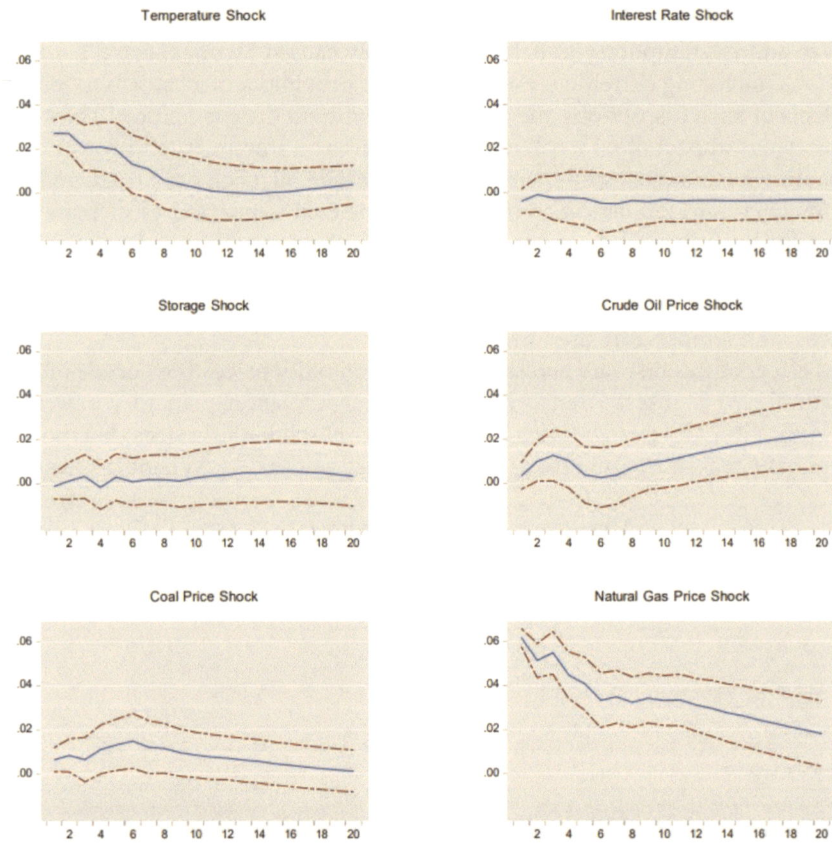

Figure 2.
Responses of natural gas prices to shock from the other variables. Notes: response is to Cholesky one s.d. innovations 2 s.e.

substantial impact on the price of natural gas through demand for heating or cooling in residential and business buildings. The effect is significant and declines over time but can last for as long as ten weeks, which indicates that temperature shocks have a short-term impact on natural gas prices. Short-term interest rate innovation has no significant effect on gas prices and crude oil and coal prices, which is contrary to the observed a priori assumption.

Innovations from storage do not also affect gas price movement. This finding is consistent with the results of another empirical research by Wang et al. [2], who find that the influence of storage on natural gas prices proves to be marginal in the U.S. market. Although this research focuses on the determinants of natural gas price variability, it is worth discussing the structural response of storage to all variables, as shown in **Figure 3**. The impulse response function reveals that extraordinary temperatures lead to storage withdrawals. This phenomenon is caused by the demand for temperature-sensitive natural gas in the heating and cooling of residential and commercial facilities. Nonetheless, the response of storage flows to natural gas price shocks is insignificant, and it may be inconsistent with economic expectations. This is because an increase in natural gas prices should incentivize storage holders to withdrawal from their inventory. This economic narrative may not be the case for U.S. market. We reason that the effect of the rise in natural gas prices in response to storage withdrawal may be muted by cross-commodity effects. Storage withdrawals occur in response to shocks in coal prices as shown in Appendix A. It could be explained by the cross-commodity relationship emphasized earlier. An increase in coal prices implies high demand; therefore, electricity producers who can switch their input move to natural gas, which affects storage levels through withdrawals. Since operators are switching from a higher price (coal) to a lower priced input (natural gas); the net change in storage may not necessarily lead to a significant price movement in natural gas. Consequently, innovations in storage leads to natural gas price stability.

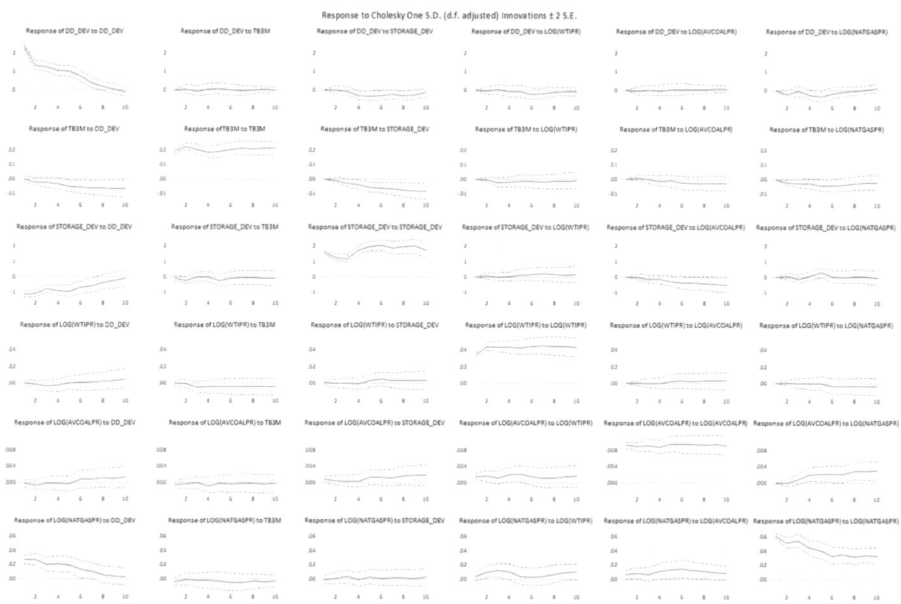

Figure 3.
Impulse response functions for all variables.

Coal prices only affect natural gas prices with a delay, but its effect is substantial and remains stable over time and gradually declines. The strong inter-dependency between coal and natural gas can be attributed to the energy mix of the U.S. electricity generation industry. According to the U.S. Energy Information Administration (EIA), about 35.1% and 27.4% of total electricity generation output were from coal and natural gas, respectively in 2018. This brings about fuel competition among energy producers, which may induce a positive cross-price elasticity of these two commodities. Consequently, a rise in the price of coal implies an increase in the demand for natural gas and, therefore a resulting price increase.

The derived structural response of natural gas prices to crude oil and coal prices reveals significant inter-dependencies among energy commodities. The price of gas responds positively to innovations in both oil and coal prices. However, the pattern with which oil and coal influence gas prices is fundamentally different. Oil price shocks impact natural gas prices instantly, causing the cyclical price movement in natural gas in the short-term, but later establish a long-run relationship after that. This long-run relationship between natural gas and oil prices is plausible as there exists a physical link in their direct substitution relation in the residential heating sector.

4.2 Forecast error variance decomposition

After identifying the structural dynamics in the natural gas market, it is appropriate to investigate the fundamental influences of each variable on natural gas price movement. To this end a forecast error variance decomposition was performed using the results of the estimated structural VAR model. The variance decomposition splits the forecast error variance of each endogenous variable, at different forecast horizons, into the components due to each of the shocks. This provides information on their relative importance. The contributions of innovations in variables calculated through the moving average representation of the VAR model are presented in **Table 3**.

Temperature and storage deviations unexpectedly have no explanatory power on the price formulation of natural gas. However, the forecast errors of gas prices can be explained more precisely by developments related to coal and oil markets. Coal price variations have the most impact on natural gas price variability, with maximum explanatory power in short-term horizons (4–12 weeks). Long-term natural gas price

Period	Temperature	Interest rate	Storage	Crude oil price	Coal price	Natural gas price
1	0.00	0.37	0.02	16.88	72.60	10.13
2	0.00	0.18	0.02	16.99	72.79	10.03
4	0.00	0.14	0.02	17.04	72.84	9.97
12	0.00	0.32	0.01	17.00	72.83	9.84
26	0.00	0.37	0.01	17.19	72.68	9.75
52	0.00	0.40	0.01	17.37	72.50	9.72

Notes: values are expressed in percentages.

Table 3.
Forecast error variance decomposition for natural gas price.

development (up to 52 weeks) is heavily affected by variations in oil prices. Coal and crude oil prices account for about 90% of the gas price variance with a forecast horizon of half a year.

4.3 Markov switching

Having identified coal prices to have the most explanatory power on natural gas price development, a Markov process was ran to ascertain gas price regime changes as explained by coal price variability. Coal prices are the first-differenced to obtain stationarity, while natural gas price series are stationary at the level.[4] A three-state Markov switching process was identified for the interaction between natural gas and coal prices, as presented in **Table 4**.

As shown in **Table 4**, Regime 1 shows a robust positive co-movement between gas and coal prices. Regime 2 (with a relatively moderate co-movement) does show a discernable switch from Regime 1 as observed in the coefficients. Regime 3, however, shows a weak co-movement. Explicitly, Regime 1 dominates the Markov process with a constantly expected duration of 56 weeks which is longer than the 24 weeks duration for Regime 2. The probability of remaining in that regime on any given day of 0.982 is the highest.

The results mean that when there is a strong co-movement between natural gas and coal prices, one can expect it to last for a long as a year which is a high probability. Displayed in **Figure 4** are the filtered switching probabilities for the Dominant Regime 1.

Variable	Coefficient	Std. error	z-Statistic	Probability
Regime 1				
C	2.72055	0.03006	90.50647	0.00000
D(COALPR)	0.20023	0.06242	3.20756	0.00130
Regime 2				
C	3.85925	0.04661	82.80654	0.00000
D(COALPR)	0.20015	0.10493	1.90754	0.05650
Regime 3				
C	4.96656	0.10550	47.07560	0.00000
D(COALPR)	0.09762	0.18518	0.52716	0.59810
Common				
LOG(SIGMA)	−0.90061	0.03500	−25.73061	0.00000

Notes: Dependent variable is the natural gas price.

Table 4.
Three state Markov switching regression for gas and coal price interactions.

[4] T-stat (gas price) = −3.546417; T-stat (difference coal price) = −20.81579.

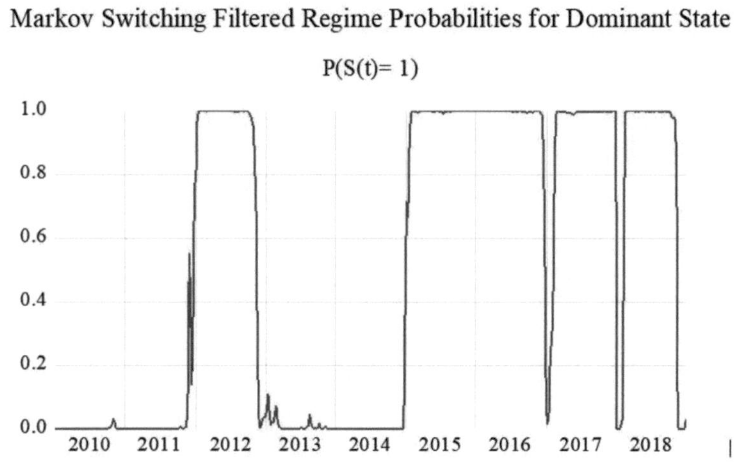

Figure 4.
Markov switching probabilities for Regime 1.

5. Conclusion

Thus far, this paper has investigated the dynamic interaction between the fundamental variables within the U.S. natural gas market. The analysis is conducted with structural VAR and Markov switching models. This approach allows us to disentangle the effects of different fundamental influences on natural gas prices. The empirical results reveal that abnormal temperatures affect natural gas prices in the short term. However, the price development of natural gas is closely tied to crude oil and coal prices in the long term, indicating the high importance of cross-commodity effects.

The specific contribution of the main fundamental drivers of gas price formation within the period under study was explicitly analyzed. Results from this study revealed that coal prices account for about 73% of the price variability of natural gas prices with a persistently strong regime switch. The results of this research are relevant to storage operators, electricity producers, and speculative investors in strategy formation and operations. As the results of this research are compelling, it cannot be generalized as natural gas prices are mainly determined by regional supply, demand, and temperature variations (see, for example, [3]). The approach provides an innovative framework for further research on more specific economic mechanisms within the U.S. gas market. Additionally, this model can be used to study other regions within the U.S. The current application is still restricted by the limited data on certain variables used in this research. A future attempt will be made to generalize the findings for the entire U.S. gas market when more data become available.

This chapter was initially published as a preprint: Nkwantabisa D. Determinants of Natural Gas Prices in the United States—A Structural VAR Approach [Internet]. SSRN Electronic Journal. Elsevier BV; 2021. Available from: http://dx.doi.org/10.2139/ssrn.3905560

JEL classifications

C32, E32, G13, Q40

Author details

Dixon Domfeh[1,2]

1 Jack Welch College of Business and Technology, Sacred Heart University, Connecticut, USA

2 Department of Computer Science, Georgia Institute of Technology, Atlanta, GA, USA

*Address all correspondence to: dnkwantabisa3@gatech.edu

References

[1] Xu Gong YL, Wang X. Dynamic volatility spillovers across oil and natural gas futures markets based on a time-varying spillover method. International Review of Financial Analysis. 2021;**76** (101790)

[2] Wang T, Zhang D, Broadstock DC. Financialization, fundamentals, and the time-varying determinants of us natural gas prices. Energy Economics. 2019;**80**: 707-719

[3] Zhang D, Ji Q. Further evidence on the debate of oil-gas price decoupling: A long memory approach. Energy Policy. 2018;**113**:68-75

[4] Wang T, Zhang D, Ji Q, Shi X. Market reforms and determinants of import natural gas prices in China. Energy Policy. 2020;**196**

[5] Nick S, Thoenes S. What drives natural gas prices? — A structural var approach. Energy Economics. 2014;**45**: 517-527

[6] Hailemariam A, Smyth R. What drive volatility in natural gas prices? Energy Economics. 2019;**80**:731-742

[7] Hartley PR, Medlock KB III. The relationship between crude oil and natural gas prices: The role of the exchange rate. The Energy Journal. 2014; **35**(2):25-44

[8] Brown SP, Yucel MK. What drives natural gas prices? The Energy Journal. 2008;**29**(2):45-60

[9] Ramberg DJ, Parsons JE. The weak tie between natural gas and oil prices. The Energy Journal. 2012;**33**(2):13-36

[10] O. S. M., "Driving fundamentals of natural gas price in Europe." International Journal of Energy Economics and Policy. 2020;**10**(6): 318-324

[11] Jian Chai ZGZSK, Liang T, Liu Z. Rationality of natural gas prices and the determining factors in China. Emerging Markets Finance and Trade. 2019;**55**(6): 1229-1246

[12] Xiaoyi M. Weather, storage, and natural gas price dynamics: Fundamentals and volatility. Energy Economics. 2007;**29**(1):46-63

[13] Taghizadeh Hesary F, Yoshino N. Monetary policies and oil price determination: An empirical analysis. OPEC Energy Review. 2014;**38**(1):1-20

[14] Ghysels E, Marcellino M. Applied Economic Forecasting Using Time Series Methods. Amsterdam, The Netherlands: Elsevier; 2018

[15] Arango LE, Arias F, Flórez A. Determinants of commodity prices. Applied Economics. 2012;**44**(2): 135-145

[16] Frankel JA. The Effect of Monetary Policy on Real Commodity Prices. Amsterdam, The Netherlands: Elsevier; 2006

[17] Pindyck RS. Volatility in natural gas and oil markets. Journal of Energy and Development. 2004;**30**:1

[18] Brigida M. The switching relationship between natural gas and crude oil prices. Energy Economics. 2014;**43**:48-55

[19] Jadidzadeh A, Serletis A. How does the us natural gas market react to demand and supply shocks in the crude

oil market? Energy Economics. 2017;**63**: 66-74

[20] Hou C, Nguyen BH. Understanding the us natural gas market: A Markov switching var approach. Energy Economics. 2018;**75**:42-53

[21] Frankel JA. Expectations and commodity price dynamics: The overshooting model. American Journal of Agricultural Economics. 1986;**68**(2): 344-348

Chapter 5

Application of Supply Chain Strategies to Improve the Performance of Syngas Networks

Nasiru Zubairu and Mira Al Balushi

Abstract

Production and utilization of syngas have the benefits of reducing greenhouse gas emissions and improving energy security. Renewable energy can be generated from syngas by converting problematic waste products into useful fuels and can be produced onsite to reduce transmission losses and costs. However, syngas is experiencing slow investment, production, and utilization due to bottlenecks, which are rooted in its supply chain networks. To address these challenges, this study aims to explore and evaluate supply chain strategies that drive performance in syngas networks to guide stakeholders to develop and embrace supply chain initiatives for improved competitiveness, sustainability, and energy security. Procurement, production, distribution and logistics, and end-use adoption are established as the key supply chain strategies that should be embedded in syngas supply chains to improve performance. Collaboration and technology support these strategies. These strategies will address the supply chain challenges, including feedstock availability, technology limitations and distribution infrastructure inadequacies.

Keywords: syngas, supply chain networks, supply chain challenges, supply chain strategies, sustainable energy, energy security

1. Introduction

The world's population has risen to eight billion in 2022, more than three times higher than in the mid-twentieth century. The global population is projected to grow to 8.5 billion by 2030 and 9.7 billion by 2050 [1]. This rapid growth, accelerating urbanization, technological advancements and deployments, and the desire for clean energy will place immense pressure on energy supply networks [2]. To address the severe environmental and economic concerns of the 21st century, the transition to sustainable and low-carbon energy is a priority [3]. Households, governments, and supply chains demand more environmentally friendly, reliable, and affordable energy sources [4, 5]. Furthermore, globalization, increasing competition, uncertain business environments, progressive customer attitudes, and the desire for sustainable energy are driving sustainable energy supply chain networks [6].

Alternative energy sources are becoming increasingly prominent due to the negative effects of traditional fossil fuels on the environment, including climate change [7]. Syngas offers an innovative solution that integrates sustainability, flexibility, and scalability [3]. Syngas is produced from carbon biomaterials [8]. Investment and deployment of syngas have the benefits of reducing greenhouse gas and energy cost. Renewable energy can be generated from syngas by converting problematic wastes to valuable fuels and can be produced onsite to reduce transmission losses and costs [7]. According to an article in the Financial Times, some of the recent developments in syngas include a $2 billion plant in the US that will produce syngas using wind and solar power. In this plant, synthetic methane, similar in chemical structure to natural gas, can be produced with captured or atmospheric carbon dioxide (CO_2), making it carbon neutral. It can be used in existing natural gas infrastructure and contribute to decarbonizing activities that are difficult to electrify. Additionally, syngas reuses CO_2 that would otherwise be released, making it superior to natural gas in terms of emissions [9].

There are examples of practical syngas utilization in real-life, including in chemical and petrochemical industries and as synthetic natural gas (SNG) for heating and transport. Syngas serves as a vital feedstock to produce various chemicals and petrochemicals. The versatile nature of syngas allows for the synthesis of a wide range of valuable products, such as methanol, ammonia, hydrogen, and synthetic fuels [8]. A notable example is the Eastman Chemical Company's methanol production plant in Texas, USA. The plant utilizes natural gas as a feedstock to produce syngas through steam reforming. The syngas is then converted into methanol, which serves as a raw material for various chemical products, including plastics, adhesives, and solvents [10]. Additionally, syngas can be further processed to produce SNG, which can be used for heating and transportation purposes. SNG production involves the conversion of syngas into methane, resulting in a gas that closely resembles natural gas in composition and properties [6]. An example is the SNG plant in Prenzlau, Germany, operated by E.ON Bioerdgas GmbH. The plant produces SNG from biogas derived from organic waste. The produced SNG is injected into the natural gas grid and distributed for heating applications, and fuel for vehicles running on compressed natural gas [11].

Despite the benefits of syngas and their promising potential, some challenges accompany the production of syngas, including the cost of producing green hydrogen and CO_2 from biomass, as well as concerns about methane leakage and the need for broader climate change mitigation [9]. Additionally, critics and skeptics raise concerns about the negative impacts and drawbacks associated with syngas production and utilization. One primary concern is the carbon intensity of syngas, particularly when produced from fossil fuel feedstocks such as coal or natural gas [12]. While SynGas can be a substitute for natural gas, it still contributes to carbon emissions and climate change if produced from non-renewable sources [13]. Another concern is syngas production processes resulting in potential environmental pollution and adverse health effects. The gasification of coal or biomass can release pollutants such as Sulfur compounds, particulate matter, and trace elements into the atmosphere [14]. The proper management of emissions and waste by-products is crucial to mitigate these environmental and health risks [15]. Further, critics argue that syngas may divert attention and resources from investing in more established renewable energy sources, such as solar and wind power. Arguing that focusing on the development of renewable technologies and energy storage systems would yield greater long-term benefits in terms of sustainability and carbon reduction [16]. Consequently, several supply chain challenges must be addressed to facilitate the substantial adoption of syngas globally. The availability of feedstock, technological innovation, cost competitiveness, regulatory frameworks, and public acceptance are some of the challenges facing syngas supply networks [17]. However, addressing these issues can open various opportunities, such as creating integrated waste management systems, increasing regional energy independence, and decreasing greenhouse gas emissions [6].

Prior studies have established the links between supply chain strategies and energy systems performance [18]. Thus, there is potential for syngas practitioners to apply supply chain initiatives to improve performance across the entire value chain. Energy networks, including that of syngas, are extended, more complex, volatile, and more prone to various supply chain challenges [8]. Syngas supply chain network encompasses a range of interdependent activities, from feedstock sourcing to gasification, purification, and distribution [19]. Understanding the complexities of each component is critical for improving efficiency, lowering environmental consequences, and increasing economic viability [6]. Thus, analyzing unconventional energy supply chains such as syngas is important owing to unexpected risks, disruptions and vulnerabilities that affect their supply networks.

2. Syngas supply chain networks

The syngas supply chain consists of various interconnected value chains, each playing a crucial role in the production, purification, distribution, and utilization of syngas. These value chains include feedstock sourcing and pre-processing, gasification and gas purification, and syngas distribution and consumption by end-users [19]. Understanding the interactions and interdependencies among these components is essential for optimizing the efficiency and sustainability of the syngas supply chain [20].

2.1 Feedstock sourcing and pre-processing

The first process in the syngas supply chain, like most supply networks, consists of sourcing raw material, in this case, feedstock. Sourcing is the strategic selection of the materials or services organizations need to run their operations, which includes supplier selection, contract negotiation, and relationship management [21]. Feedstock acquisition involves sourcing suitable carbon-containing materials such as biomass, coal, or waste [22]. The selection of feedstock depends on factors such as availability, cost, sustainability, and local regulations [19]. Once the suitable feedstock has been sourced, the next value chain is the pre-processing, which is preparing the feedstocks to meet the gasification process requirements and may include activities such as sorting, shredding, drying, and size reduction to optimize gasification performance [20].

2.2 Gasification and purification

Once the feedstock has been pre-processed, it is ready for the gasification process, which is a critical process in syngas production, where feedstocks are converted into a mixture of carbon monoxide and hydrogen [19]. Different gasification technologies, such as entrained flow, fluidized bed, or plasma gasification, are utilized based on specific feedstock properties and process requirements [3]. Gasification efficiency, feedstock flexibility, and the control of by-products such as tar and ash are critical considerations in this stage [23]. After gasification, the raw gas undergoes purification and treatment to remove impurities and contaminants. This includes the removal of particulate matter, Sulfur compounds, nitrogen compounds, and trace elements [19]. Various gas treatment technologies, such as wet scrubbing, adsorption, and catalytic conversion, are employed to meet the required quality standards for syngas [23]. Effective purification and treatment are crucial for maximizing the energy value and minimizing the environmental impacts of the final syngas product [3].

2.3 Distribution and end-use applications

Once purified, the syngas must be efficiently transported and distributed to end users. The supply chain's distribution process connects syngas's production to the downstream entities or end-user [20]. The distribution infrastructure may involve pipelines, compressed gas transportation systems, or storage facilities [24]. Distance, logistics, and safety considerations are vital in designing an effective distribution network [19]. Therefore, collaborations with existing natural gas distribution systems and the development of dedicated syngas pipelines can enhance the reach and accessibility of syngas as a viable energy source [20]. The syngas supply chain's final stage involves utilizing syngas in various applications. These applications include power generation, heating, and industrial processes [3]. Additionally, syngas can be directly used as fuel in gas turbines or boilers, blended with natural gas, or converted into value-added chemicals for further processing [19]. The versatility of syngas applications offers significant opportunities for energy diversification and decarbonization across sectors [23].

Supply chain technologies play a vital role in each of the syngas value chains discussed. Advanced technologies are used in each stage of the supply chain, making technology providers crucial stakeholders in the syngas supply systems. Technological advances contribute to syngas production's efficiency, scalability, and reliability [6]. Other stakeholders in the syngas supply chain include feedstock providers, gas treatment companies, energy producers, regulatory bodies, and policymakers [25]. Feedstock providers play a crucial role in ensuring the availability and quality of suitable feedstocks for syngas production [3]. The feedstock suppliers include biomass suppliers, coal mines, waste management companies, and agricultural producers [22]. Collaboration with feedstock providers is essential for securing a consistent and sustainable supply, exploring novel feedstock sources, and adhering to responsible sourcing practices [26]. Gas treatment companies ensure the removal of impurities and contaminants, leading to a high-quality syngas product [27]. Energy producers, such as power generation companies or industrial facilities, use syngas as fuel. They play a pivotal role in adopting syngas-based technologies, optimizing combustion processes, and integrating syngas into their energy portfolios. Regulatory bodies and policymakers play a crucial role in shaping the regulatory framework, incentives, and guidelines that govern the syngas supply chain [28]. Collaboration between stakeholders and regulatory bodies is essential to align industry practices with sustainability goals, encourage research and development, and ensure a level playing field for syngas within the energy market [25]. Understanding the stakeholder and value chain requirements and responsibilities is crucial for fostering collaboration, identifying potential bottlenecks, and promoting sustainable practices within the syngas supply chain networks [6].

3. Supply chain challenges associated with syngas networks

Syngas is an alternative to traditional fossil fuels and offers reduced greenhouse gas emissions and enhanced sustainability [8]. However, the successful integration of syngas into the energy mix relies on addressing the unique challenges within its supply

Challenges	Causes	Authors
Feedstock Availability	• Variability of feedstock composition and characteristics • Competition for feedstock • Sustainability and environmental considerations	[12, 13, 23–26]
Gasification Technology Limitations	• Technology maturity and scalability • Flexibility in feedstock utilization	[13, 16, 27, 29]
Distribution Infrastructure Requirements	• Pipeline infrastructure • Storage and compression	[3, 7, 8]
End-use Applications	• Technological integration • Market acceptance and demand	[7, 15, 19, 22]

Table 1.
Supply chain challenges facing syngas networks.

chain networks. These challenges include feedstock availability, gasification technology limitations, distribution infrastructure requirements, and end-use applications [29–31]. Understanding and mitigating these challenges is crucial for ensuring syngas efficiency, sustainability, and widespread adoption as a viable energy source [30]. **Table 1** and **Figure 1** present the supply chain challenges associated with syngas networks.

3.1 Feedstock availability

One of the primary challenges in the syngas supply chain is ensuring a reliable and sustainable feedstock supply (**Table 2**). The availability and accessibility of feedstock

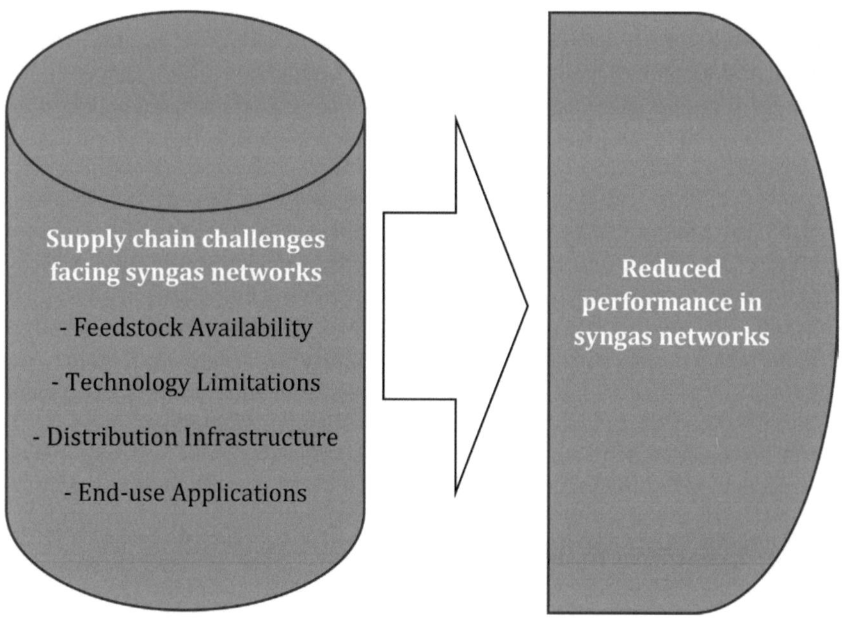

Figure 1.
Supply chain challenges affecting syngas networks.

Strategies	Initiatives	Authors
Procurement	• Sourcing and procurement management • Strategic partnerships and alliances	[32–36]
Production	• Technology selection and optimization • Process integration and efficiency	[33, 36–39]
Distribution and Logistics	• Infrastructure development and optimization • Supply chain visibility and collaboration	[23, 40–42]
End-use Adoption	• Market development and demand creation • Technological integration and innovation	[39, 43–46]

Table 2.
Supply chain strategies that improve performance in syngas networks.

significantly impact the feasibility and cost-effectiveness of syngas production [20]. Feedstock sources include coal, biomass, waste materials, and natural gas. Studies have highlighted the importance of feedstock diversification to mitigate supply chain risks and promote sustainability [31]. Feedstock availability has been identified as a significant challenge in the syngas supply chain due to the variability of feedstock composition and characteristics, competition for feedstock, and environmental considerations [19, 31, 32]. The need for developing alternative feedstocks and evaluating their suitability for syngas production is critical. Ensuring a consistent supply of feedstock and exploring novel sources are crucial to addressing feedstock shortages [19, 31, 47]. Syngas production relies on various feedstocks, such as biomass, coal, or waste materials [47]. However, the composition and characteristics of these feedstocks can vary significantly, posing challenges to the gasification process. Variations in moisture content, ash content, and feedstock quality can impact gasification efficiency and result in varying syngas compositions [19]. Ensuring a consistent and reliable feedstock supply requires strategic activities such as blending, pre-processing techniques, and advanced feedstock characterization methods [20]. The availability and accessibility of feedstocks for syngas production can be limited due to competition with other industries, such as agriculture, waste management, or traditional energy production [32]. This competition can result in price fluctuations and supply constraints [12]. Thus, collaborative efforts among stakeholders, including feedstock providers, policymakers, and researchers, are essential to identify alternative feedstock sources, developing sustainable feedstock management practices, and establishing long-term supply agreements [20]. Additionally, syngas production can support curtailing greenhouse gas emissions and promote environmental sustainability [30]. However, the sustainability of feedstock sourcing and production processes must be carefully assessed [47]. Biomass feedstocks, for example, require attention to land use, deforestation, and agricultural practices. Balancing feedstock production's environmental impacts with syngas networks' sustainability objectives requires close collaboration among stakeholders and adherence to strict sustainability standards [32].

3.2 Gasification technology limitations

Gasification is a key process in syngas production, converting feedstock into a mixture of hydrogen and carbon monoxide. However, gasification technologies face certain limitations that affect the efficiency and reliability of syngas production [23].

The limitation of gasification technology is another challenge facing the syngas supply chain [20, 23, 40]. Research efforts have focused on improving gasification technologies to overcome these limitations. Innovative gasification methods and reactor designs are explored by extant studies to enhance process efficiency and address feedstock variability [20, 23]. More efforts are required for continued advancements in gasification technologies to optimize syngas production. Gasification technologies used in syngas production continuously evolve, and their maturity and scalability can pose challenges [20]. Many gasification processes are still in the research and development phase or are limited to small-scale applications. Scaling up these technologies for commercial production requires significant investment, technical advancements, and demonstration projects [33]. Therefore, collaborative efforts between technology providers, researchers, and investors can drive innovation, improve gasification efficiency, and enhance the scalability of gasification technologies [48]. Additionally, different gasification technologies exhibit varying levels of flexibility in feedstock utilization [40]. Some technologies are optimized for specific feedstocks, limiting the ability to switch or adapt to changing market conditions [20]. Enhancing the flexibility of gasification technologies through process modifications, catalyst development, or hybrid systems can improve feedstock utilization, increase operational efficiency, and reduce feedstock-related risks [23].

3.3 Distribution infrastructure requirements

Transportation is required at every value stream. Effective distribution infrastructure is essential for transporting syngas from production facilities to end-use applications. Challenges in this area include pipeline connectivity, storage capabilities, and safety considerations [3]. Consequently, establishing an efficient and robust distribution network requires significant investments and coordination among various stakeholders [8]. Studies are conducted to optimize distribution infrastructure to enhance syngas supply chain performance. Studies by [3, 7, 8] analyze the design and operational aspects of syngas pipeline networks to minimize transportation costs and ensure reliable supply. However, understanding the infrastructure requirements and developing innovative solutions are critical to improving cost and reliability [7]. Efficient and reliable distribution infrastructure is crucial for transporting syngas from production facilities to end-use applications. However, the existing natural gas pipeline infrastructure may not be suitable for syngas due to differences in composition and impurity content [8]. Developing dedicated syngas pipelines or retrofitting existing infrastructure requires substantial investments and collaboration between syngas producers, pipeline operators, and regulatory authorities, creating a challenge for syngas supply chains [3]. There are also challenges associated with the storage of syngas. The storage and compression of syngas are essential for ensuring a consistent supply and accommodating fluctuations in demand [8]. However, syngas has different properties than natural gas, affecting storage and compression requirements [7]. Therefore, developing storage technologies suitable for syngas, such as high-pressure tanks or gas holders, and optimizing compression processes can enhance the flexibility and reliability of syngas distribution networks [8].

3.4 End-use applications

The successful adoption of syngas relies on identifying and developing viable end-use applications. End-use applications, such as technology integration and market acceptance and demand are challenges affecting syngas supply chain performance [7, 22, 29]. Each application has unique requirements and regulations that must be considered in the syngas supply chain [26]. The feasibility and economic viability of different end-use applications of syngas have been explored to understand and optimize its utilization in different sectors, which is crucial for achieving market acceptance [29]. Integrating syngas into existing energy infrastructure and end-use applications can present technological challenges [7]. Syngas utilization requires modifications or replacement of combustion systems, such as boilers or gas turbines, to ensure compatibility and optimal performance [22]. Collaborations between syngas producers, equipment manufacturers, and end-users are crucial for identifying technological requirements, conducting feasibility studies, and implementing syngas-based solutions [30]. The successful adoption of syngas relies on market acceptance and demand for syngas-based products and applications [26]. Building awareness, educating stakeholders, and showcasing the advantages of syngas in terms of emissions reduction, energy security, and cost competitiveness is essential to enhancing performance across syngas supply chains [29]. Establishing favorable policies, financial incentives, and market mechanisms can drive market demand and facilitate the significant integration of syngas into the global energy mix [7].

4. Linkages between supply chain strategies and energy networks

The transition towards sustainable and low-carbon energy systems requires a deep understanding of the linkages between supply chain strategies and energy networks [49]. Supply chain strategies encompass a range of strategic initiatives deployed in procurement, production, distribution, and logistics [18]. Similarly, energy networks involve energy generation, transmission, and distribution [50]. Understanding the linkages between supply chain strategies and energy networks, including interdependencies, challenges, and opportunities, is crucial for optimizing energy systems, enhancing supply chain efficiency, and achieving sustainability goals [51]. Interdependencies between supply chain strategies and energy networks include procurement, which is energy sourcing, production as energy generation, and distribution encompassing energy transmission. Procurement strategies are vital in determining the energy sources utilized within an energy network [52]. Strategic decisions regarding the sourcing of renewable energy, such as solar, wind, or hydro, versus conventional fossil fuels directly impact the energy network's environmental sustainability and carbon footprint [51].

Collaborative efforts between supply chain managers and energy network operators are essential for aligning procurement strategies with sustainability objectives, exploring renewable energy options, and ensuring a diversified and reliable energy supply [49]. Production strategies within supply chains are closely linked to energy generation methods [53]. The energy requirements for manufacturing processes, such as heating, cooling, or powering machinery, influence the energy generation choices made by organizations [7]. Optimization of production processes, adoption of energy-efficient technologies, and implementation of onsite renewable energy generation can enhance the energy performance of supply chains [51]. Collaborations between supply chain managers and energy providers can enable the adoption of cleaner and more sustainable energy generation methods, reducing environmental impacts and improving energy efficiency [50]. Additionally, the efficient distribution of goods within supply chains relies on reliable and resilient energy transmission networks [54]. Energy transmission infrastructure, including power lines, substations, and grid systems, enables the transportation of electricity to end-users [7]. Supply chain managers must consider energy transmission networks' availability, reliability, and capacity to ensure uninterrupted operations [50]. Thus, collaborations between supply chain managers, energy transmission companies, and grid operators are essential for optimizing distribution processes, managing energy demand, and enhancing the resilience of energy networks [50].

Many challenges can be addressed by aligning supply chain strategies and energy networks, such as the lack of integration and collaboration in the network, uncertainty in energy supply and demand, and cost considerations and financial implications. One of the critical challenges in aligning supply chain strategies and energy networks is the lack of integration and collaboration between stakeholders [50]. Traditionally, energy supply chain and energy management have been treated as separate organizational functions. Siloed decision-making processes and fragmented approaches hinder the identification of synergies and optimization opportunities [37]. Overcoming this challenge requires enhanced communication, cross-functional collaboration, and the establishment of shared goals and metrics across the supply chain and energy management functions [34]. Consequently, the dynamic nature of energy markets and the increasing integration of renewable energy sources introduce uncertainty in both energy supply and demand [37]. Fluctuations in renewable energy generation due to weather conditions and variable energy demand patterns pose challenges for supply chain planning and operations [37]. Supply chain strategies need to be flexible and adaptive to accommodate these uncertainties. Therefore, utilizing advanced forecasting techniques, real-time data analytics, and demand-response mechanisms can help mitigate the impacts of energy supply and demand uncertainties on supply chain operations [51]. Integrating sustainable energy solutions into supply chain strategies often involves additional costs and other financial implications [34]. Investments in renewable energy infrastructure, energy-efficient technologies, and energy storage systems may require significant upfront capital. The financial viability and return on investment of such initiatives need to be carefully assessed [35]. Collaborations between supply chain managers and finance departments are crucial for evaluating the cost-benefit trade-offs, identifying financial incentives or subsidies, and developing business models that align supply chain strategies with sustainable energy objectives [50].

Overcoming challenges associated with linking supply chain strategies with energy networks presents opportunities for synergy and optimization through energy-efficient supply chain design, demand-side management, and renewable energy integration and offsetting [36]. Supply chain network design is critical in optimizing energy consumption and reducing environmental impacts [38]. Designing supply chain networks with energy efficiency in mind involves considerations such as facility location, transportation modes, packaging, and inventory management. Implementing strategies such as consolidation of shipments, route optimization, and green logistics practices can minimize energy consumption and greenhouse gas emissions [36]. Therefore, the collaboration between supply chain managers, logistics providers, and energy experts can drive the integration of energy-efficient practices into supply chain design [43]. Demand-side management strategies assist in managing and optimizing energy demand to reduce peak loads, enhance energy efficiency, and support grid stability [39]. Supply chain managers can contribute to demand-side management by aligning production schedules, transportation activities, and facility operations with energy demand patterns [41]. By strategically adjusting energy-intensive operations during off-peak hours or implementing load-shifting techniques, energy supply chains can support a more balanced and efficient use of energy resources [38]. Energy supply chains can contribute to expanding renewable energy generation by integrating onsite systems or participating in purchase agreements [43]. By generating renewable energy or offsetting their energy consumption through renewable energy certificates, energy supply chains can support the growth of clean energy sources and reduce their carbon footprint [38]. Thus, collaborations between supply chain managers, renewable energy providers, and sustainability experts are crucial for identifying viable renewable energy integration opportunities and implementing strategies to improve performance [39].

5. Supply chain strategies as drivers of performance in syngas networks

Supply chain strategies are critical in optimizing operations and driving performance in syngas networks [49]. The successful integration of syngas into the energy mix requires robust supply chain strategies encompassing procurement, production, distribution, and end-use applications [50]. **Table 2** and **Figure 2** present the supply chain strategies that can be adopted to address the challenges affecting syngas networks. Procurement and sourcing strategies are identified as some of the most effective supply chain strategies [42, 44, 45, 52, 54]. These strategies include feedback sourcing, management, strategic partnerships, and alliances [45]. The choice of feedstocks and their efficient procurement directly impacts the performance of syngas networks. Effective procurement strategies involve identifying reliable and sustainable sources, establishing long-term supply agreements, and ensuring feedstock quality and consistency [54]. Collaborative efforts between syngas producers, feedstock suppliers, and regulatory bodies drive the development of sustainable feedstock management practices, such as feedstock blending, waste-to-energy conversion, or biomass cultivation, which enhance the reliability and sustainability of feedstock

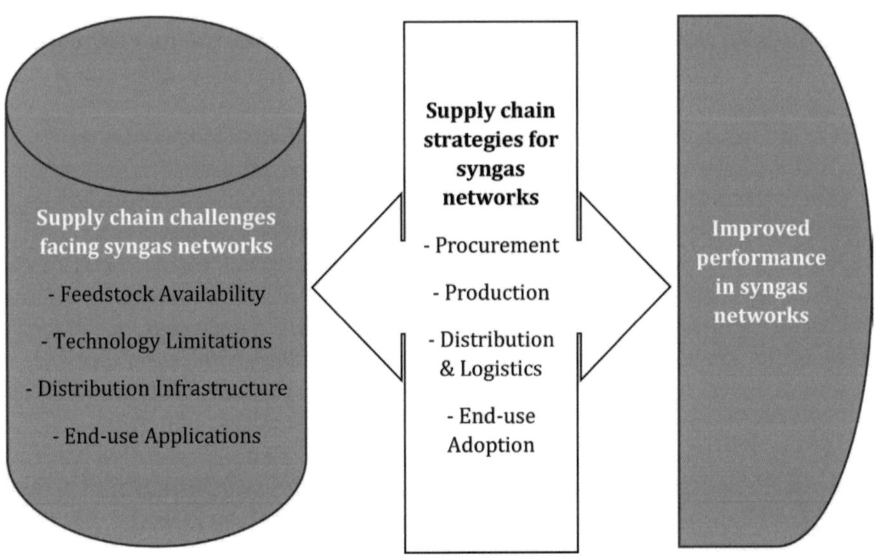

Figure 2.
Supply chain strategies for improving performance in syngas networks.

supply [42]. Further, strategic partnerships and alliances with suppliers, technology providers, and research institutions can unlock synergies and drive performance improvements in syngas networks [52]. Collaborative relationships enable knowledge sharing, technology transfer, and access to innovative solutions [44]. Establishing strategic partnerships can also enhance supply chain resilience by mitigating supply disruptions, improving access to critical resources, and enabling joint research and development efforts for continuous improvement and optimization [42].

Production strategies are also effective in deriving supply chain performance [45, 46, 54, 55]. These strategies include technology selection, optimization, process integration, and efficiency improvement [56]. Gasification technology selection is a crucial aspect of production strategies in syngas networks [55]. Different gasification technologies have varying efficiency, scalability, and feedstock compatibility [56]. Evaluating and selecting the most suitable gasification technology based on feedstock characteristics, network requirements, and sustainability objectives is essential [54]. Optimizing gasification processes, including reactor design, operating conditions, and catalyst selection, can enhance syngas production efficiency, reduce emissions, and improve overall network performance [46]. Integrating syngas production processes with other industrial operations or energy systems can unlock efficiency gains and resource utilization opportunities [54]. Waste heat recovery, cogeneration, or integration with combined heat and power systems can enhance energy efficiency, reduce energy costs, and improve overall process performance [46]. Further, process optimization techniques, such as advanced control systems, real-time monitoring, and predictive analytics, enable continuous improvement and enhance the performance of syngas production processes [45].

Distribution and logistics strategies are utilized in supply chains to boost performance [30, 53, 57, 58]. These strategies include infrastructure development, optimization, supply chain visibility, and collaboration [30]. Syngas's efficient distribution and logistics require a well-designed infrastructure and optimized transportation networks [53]. Developing dedicated pipelines or retrofitting existing infrastructure ensures the safe and reliable transmission of syngas [58]. Optimal pipeline routing, capacity planning, and compression strategies are crucial to minimize energy losses and maximize network efficiency [57]. Additionally, the integration of advanced monitoring and control systems enables real-time optimization. It improves the reliability and responsiveness of the distribution network [53]. Supply chain visibility and stakeholder collaboration are vital strategies for improving logistics and distribution performance in syngas networks [30]. Real-time data sharing, supply chain analytics, and collaborative platforms enhance coordination, enable demand forecasting, and optimization of transportation and inventory management [58]. Collaborations between syngas producers, distributors, and end-users can streamline operations, reduce inefficiencies, and enhance overall supply chain performance [57].

End-use adoption strategies, such as market development, demand creation, technology integration, and innovation, can enhance supply chains [59]. Practical strategies for end-use applications are crucial for creating and expanding market demand for syngas-based products and services [60]. Syngas can be utilized in various sectors, such as power generation, chemical manufacturing, or transportation [59]. Market development efforts include identifying target industries, conducting market assessments, and developing tailored marketing and sales strategies [55]. Thus, the collaboration between syngas producers, equipment manufacturers, and industry associations is essential for promoting syngas deployment, and showcasing its benefits to drive market adoption [56]. However, the integration of syngas into end-use applications requires the utilization of advanced technologies and innovation [61]. Collaborations with technology providers, research institutions, and end-users facilitate the identification of technological tools, developing customized solutions to demonstrate syngas feasibility [56]. Technological integration strategies encompass retrofitting existing equipment, developing new conversion technologies, and exploring synergies with emerging trends such as carbon capture and utilization [55]. Embracing technological innovation enhances syngas-based end-use applications' performance, efficiency, and sustainability [61].

Once the relevant supply chain strategies are implemented, developing related key performance indicators and metrics is crucial for measuring and evaluating the performance of syngas networks [62]. The key performance indicators may include energy efficiency, carbon emissions, feedstock utilization, network reliability, and customer satisfaction [63]. Performance measurement provides insights into the effectiveness of supply chain strategies, identifies areas for improvement, and enables benchmarking against industry standards [64]. Continuous monitoring and analysis of performance metrics enable proactive decision-making and the identification of opportunities for further optimization [65]. Additionally, establishing a continuous improvement and innovation culture is essential for driving performance in syngas networks [62]. Supply chain managers should encourage experimentation, foster a learning environment, and facilitate knowledge sharing among stakeholders [64]. Embracing technological advancements, industry best practices, and sustainability practices enable the identification of innovative strategies, process improvements, and adoption of emerging technologies that enhance overall network performance [66].

6. Directions for future research and development on syngas supply chains

Syngas supply chain network is a complex and evolving area that requires continuous research and development to address emerging challenges, optimize operations, and drive performance and sustainability [19]. Future studies on syngas networks should explore advanced and innovative supply chain initiatives, such as integrating renewable energy sources into the networks, supply chain digitalization, and circular economy approaches [67]. Integrating renewable energy sources into syngas supply chains represents a promising avenue for future research. Renewable energy technologies, such as solar, wind, and biomass, can be utilized for syngas production, reducing the reliance on fossil fuels, and mitigating environmental impacts [68]. Research efforts should focus on exploring advanced gasification technologies, optimizing the integration of renewable energy sources with syngas production processes, and assessing such integration's environmental, social, and economic benefits [69]. Investigating the role of energy storage systems, innovative grid technologies, and demand response mechanisms in enabling efficient and reliable syngas production from renewable sources would contribute to developing sustainable syngas supply chains [67]. The digital transformation of supply chains presents numerous opportunities for enhancing syngas supply chains' efficiency, transparency, and resilience [70]. Future studies should consider investigating the application of emerging technologies, such as blockchain, internet of things, artificial intelligence, and data analytics, to improve supply chain visibility, optimize inventory management, enable real-time monitoring of processes, and facilitate seamless collaboration among stakeholders [71]. Studies that will integrate digital platforms and predictive modeling techniques can support demand forecasting, risk management, and decision-making processes in syngas supply chains [72]. Research on the implementation challenges, data privacy and security considerations, and the potential for scaling up digital solutions in the context of syngas supply chains are lacking [73]. Adopting circular economy principles in syngas supply chains can promote resource efficiency, waste reduction, and the valorization of by-products [74]. Future research should explore strategies for waste-to-energy conversion, using carbon capture and utilization technologies and developing symbiotic relationships between syngas producers and other industries [75]. Investigating the economic feasibility, environmental impacts, and regulatory frameworks surrounding the integration of circular economy approaches in supply chains is essential [76]. Future research should explore the potential for industrial symbiosis and identify opportunities to co-utilize waste streams or by-products from different sectors to contribute to developing sustainable syngas supply chains [77].

7. Conclusions

This chapter has explored various aspects of syngas supply chains, including the various value chains across the supply networks, challenges, linkages with energy networks, supply chain strategies, and proposed future research directions. The study of syngas supply chains is critical in the quest for sustainable energy solutions and the transition towards a low-carbon future [19]. Syngas supply chains are complex networks involving feedstock sourcing, production processes, distribution, and end-use adoptions [3]. The study has identified and analyzed the stakeholders involved, the flow of materials and information, and the interdependencies among different value chains across syngas networks [19]. The analysis revealed the complexities and opportunities within the syngas networks [20]. The challenges associated with syngas supply chains are multifaceted and require a strategic approach for effective management [19]. These challenges include feedstock availability, infrastructure development, and market adoption [20]. The challenges require innovative solutions and stakeholder collaborations to ensure a reliable and sustainable syngas supply [7]. The linkages between supply chain strategies and energy networks have been highlighted as crucial for optimizing syngas supply chains. The application of supply chain strategies on syngas networks will enable participants to enhance operational efficiency, promote renewable energy integration, and contribute to the overall stability and security of energy systems [51]. Collaboration, market development efforts, technological integration, and performance measurement are essential strategies for enhancing the performance of syngas networks [49]. Supply chain strategies play a significant role in driving performance in syngas networks. Effective procurement, production, distribution, and end-use adoption strategies can enhance efficiency, sustainability, and network performance [42]. Feedstock sourcing, strategic partnerships, infrastructure development, supply chain visibility, market development, technological integration, and continuous improvement initiatives are vital for achieving optimal performance [56]. There are several directions for research development in syngas supply chains. This includes integrating renewable energy sources, such as solar, wind, and biomass, which promises to reduce reliance on fossil fuels and mitigate environmental impacts [67]. Digitalization of supply chains offers further opportunities for improving visibility, efficiency, and collaboration among stakeholders [70]. Adopting circular economy principles can enhance resource efficiency, waste reduction, and the valorization of by-products [74].

Acknowledgements

The research leading to these results has received funding from the Ministry of Higher Education, Research and Innovation (MoHERI) of the Sultanate of Oman under the Block Funding Program. MoHERI Block Funding Agreement No MoHERI/BFP/MUSCATUNIVERSITY/01/2020.

Author details

Nasiru Zubairu* and Mira Al Balushi
Faculty of Transport and Logistics, Muscat University, Muscat, Oman

*Address all correspondence to: nzubairu@muscatuniversity.edu.om

References

[1] United Nations. World Population Prospects 2022. 2022. Available from: https://www.un.org/development/desa/pd/sites/www.un.org.development.desa.pd/files/wpp2022_summary_of_results.pdf. [Accessed: April 30, 2023].

[2] Hrnjica B, Mehr AD. Energy demand forecasting using deep learning. In: Al-Turjman F, editor. Smart Cities Performability, Cognition, & Security. EAI/Springer Innovations in Communication and Computing. Cham: Springer; 2020. DOI: 10.1007/978-3-030-14718-1_4

[3] Ahmadvand S, Sowlati T. A robust optimization model for tactical planning of the forest-based biomass supply chain for syngas production. Computers and Chemical Engineering. 2022;**159**:107693

[4] IEA. Renewables 2021: Analysis and forecasts to 2026, Available from: https://www.iea.org/reports/renewables-2021. 2021. [Accessed: April, 2023].

[5] Asif T, Muneer T. Energy supply, its demand and security issues for developed and emerging economies. Renewable and Sustainable Energy Reviews. 2007;**11**(7):1388-1413. DOI: 10.1016/j.rser.2005.12.004

[6] Ali A, Shahbaz M, Shahzad K, Inayat M, Naqvi S, Mahpudz A, et al. Polygeneration syngas and power from date palm waste steam gasification through an Aspen plus process modeling. Fuel. 2023;**332**:126120

[7] Lo S, How B, Teng S, Lam H, Lim C, Rhamdhani M, et al. Stochastic techno-economic evaluation model for biomass supply chain: A biomass gasification case study with supply chain uncertainties. Renewable and Sustainable Energy Reviews. 2021;**152**:111644

[8] Marufuzzaman M, Li X, Yu F, Zhou F. Supply chain design and Management for Syngas Production. ACS Sustainable Chemistry Engineering. 2016;**4**:890-900

[9] Hook L, Tani S. Total and Belgian energy start-up Plan US Synthetic Gas Investment [Internet]. Financial Times. 2023. Available from: https://www.google.com/url?sa=t&rct=j&q=&esrc=s&source=web&cd=&cad=rja&uact=8&ved=2ahUKEwj-v6Gh5aL_AhWhS_EDHV80BjgQFnoECAsQAQ&url=https%3A%2F%2Fwww.ft.com%2Fcontent%2F2a61de09-7a2f-415b-8050-e402c58087b7&usg=AOvVaw1E-WSn1tGrJxsA9W2FcgVi [Accessed: June 1, 2023]

[10] Eastman Chemical Company. Methanol Manufacturing [Internet]. 2023. Available from: https://www.eastman.com/Company/Businesses/Chemical_Intermediates/Methanol_Manufacturing/Pages/default.aspx [Accessed: June 17, 2023]

[11] E.ON Bioerdgas GmbH. SNG Plant Prenzlau [Internet]. 2023. Available from: https://www.eon.com/en/business-customers/biogas/sng-plant-prenzlau.html [Accessed: June 17, 2023]

[12] Piacente E, Chiesa P, Consonni S. Assessment of coal and biomass to SNG chains with CCS. Applied Energy. 2021;**293**:116894

[13] Rana R, Anwer MA, Usman M, Kumar S, Ding S. Gasification technologies for syngas production from coal: A critical review. Journal of Cleaner Production. 2021;**291**:125927

[14] Samolada MC, Zabaniotou AA, Skoulou VK. Environmental impact assessment and exergy analysis of syngas production from lignocellulosic biomass. Journal of Cleaner Production. 2020;**243**:118686

[15] Purwanto P, Jayanti DS, Mel M, Nugroho Y. Environmental impact assessment of syngas production using biomass gasification for power generation: A case study in Indonesia. Journal of Environmental Management. 2020;**259**:110043

[16] Bhattacharya S, Pakrashi V, Chattopadhyay S. State-of-the-art of sustainable gasification: An overview. Journal of Cleaner Production. 2021;**311**:127414

[17] Munasinghe PC, Khanal SK. Biomass-derived syngas fermentation into biofuels: Opportunities and challenges. Bioresource Technology. 2010;**101**(13):5013-5022. DOI: 10.1016/j.biortech.2009.12.098

[18] Zubairu N, Dinwoodie J, Govindan K, Hunter L. Supply chain strategies as drivers of financial performance in liquefied natural gas networks. Supply Chain Management: An International Journal. 2021;**26**(5):579-591

[19] Li Q, Hu G. Supply chain design under uncertainty for advanced biofuel production based on bio-oil gasification. Energy. 2014;**74**:576-584

[20] Aseffe J, Gonzalez A, Jaen R, Lora E. The corn cob gasification-based renewable energy recovery in the life cycle environmental performance of seed-corn supply chain: An Ecuadorian case study. Renewable Energy. 2022;**163**:1523-1535

[21] Zhao Q, Xiong J, Yang J, Zhang B, Wang Y, Li X. Energy consumption prediction using an improved LSTM deep learning model for a smart grid environment. Energies. 2020;**13**(6)

[22] Sowlati T, Zabihian S. Sustainable Forest bioenergy supply chains. In: Vilpoux O, editor. Biofuels - Status and Perspective. London, UK: IntechOpen; 2022

[23] Holmgren K, Berntsson T, Andersson E, Rydberg T. The influence of biomass supply chains and by-products on the greenhouse gas emissions from gasification-based bio-SNG production systems. Energy. 2015;**90**:148-162

[24] Enoch M, Mashela PW. An overview of the energy sector in the Kingdom of Lesotho and the potential for exploiting biomass energy resources. Journal of Energy in Southern Africa. 2021;**32**(1)

[25] Ramakrishna S, Bandara M, Almoustafa R. Hydrogen and syngas production and purification technologies for carbon capture and storage in the era of net-zero carbon emissions. International Journal of Hydrogen Energy. 2023;**48**(9)

[26] Oda J, Roy R. The role of clean energy technologies in building resilient infrastructure and supporting a green recovery: A comprehensive analysis of global progress. Energy Research and Social Science. 2022;**92**

[27] Lin G, Zhang J, Luo J, Wu X, Sun Y, Jin X, et al. The impact of technical factors on the levelized cost of electricity for biomass power plants: A comprehensive sensitivity analysis. Energy Conversion and Management. 2021;**246**

[28] De Gisi S, Lofrano G, Grassi M, Notarnicola M. Characteristics and adsorption capacities of low-cost sorbents for wastewater treatment:

A review. Sustainable Materials and Technologies. 2016;**9**

[29] De Pinto A, Dell'Isola M, Errico M, Incarnato L, Acierno D. Development of a comprehensive database for municipal solid waste gasification and pyrolysis technologies. Waste Management. 2023;**140**

[30] Ahuja V, Bhatt A, Ravindran B, Yang Y, Bhatia S. A mini-review on syngas fermentation to bio-alcohols: Current status and challenges. Sustainability. 2023;**15**(4):3765

[31] Shi Z, Kong F, Zhou J, Li L, Chen J. A review of air pollution control technologies in biomass power plants. Fuel Processing Technology. 2021;**221**

[32] Li H, Zhu X, Li B, Yan J, Liao Q. Upgrading syngas-derived bio-oil over an industrial Ni–Cu-based catalyst for the production of high-quality gasoline. Green Energy & Environment. 2023

[33] Zhang X, Zhang X, Li X, Ma L, Zhang X. Prediction model of urban energy consumption based on deep learning. Applied Energy. 2021;**282**

[34] Kucukvar M, Samadi H. Linking national food production to global supply chain impacts for the energy-climate challenge: The cases of the EU-27 and Turkey. Journal of Cleaner Production. 2015;**108**:395-408

[35] Sahebi I, Mosayebi A, Masoomi B, Marandi F. Modeling the enablers for blockchain technology adoption in renewable energy supply chain. Technology in Society. 2022;**68**:101871

[36] Mokadem M, Khalaf M. The contingent effect of supply chain strategies on the relationship between supply chain integration and operational performance in manufacturing context. Journal of Manufacturing Technology Management. 2023;**34**(1):147-164

[37] Gebhardt M, Beck J, Kopyto M, Spieske A. Determining requirements and challenges for a sustainable and circular electric vehicle battery supply chain: A mixed-methods approach. Sustainable Production and Consumption. 2022;**33**:203-217

[38] Iddrisu S. Sustain or perish: Can environmental supply chain strategies predict sustainable performance of manufacturing firms? A third world nation's perspective. Management of Environmental Quality: An International Journal. 2023;**34**(1):99-118

[39] Oliveria-Dias D, Marin J, Moyano-Fuentes J. Lean and agile supply chain strategies: The role of mature and emerging information technologies. The International Journal of Logistics Management. 2022;**33**(5):221-243

[40] Nanda S, Mahanta P, Moholkar VS. Techno-economic analysis of sustainable biomass-to-bioenergy systems: A review. Biomass Conversion and Biorefinery. 2022;**12**

[41] Ocicka B, Baraniecka A, Jefmanski B. Exploring supply chain collaboration for green innovations: Evidence from the high-tech industry in Poland. Energies. 2022;**15**(5):1750

[42] Bret L et al. Towards a model assessing supply chain resilience strategies. Procedia CIRP. 2021;**103**:14-19

[43] Yin Y, Liu F. Carbon emission reduction and coordination strategies for new energy vehicle closed-loop supply chain under the carbon trading policy. Complexity. 2021;**2021**:3720373

[44] Forsgren L et al. Health systems resilience in practice: A scoping review to

identify strategies for building resilience. BMC Health Services Research. 2022;**22**:1173

[45] Powell D et al. Quick-scan – Towards a strategy for responsive and resilient value chains. Procedia CIRP. 2021;**104**:1355-1360

[46] Vimal K et al. Resilience strategies to recover from the cascading ripple effect in a copper supply chain through project management. Operations Management Research. 2022;**15**:440-460

[47] Toledo M, Arriagada A, Ripoll N, Salgansky E, Mujeebu M. Hydrogen and syngas production by hybrid filtration combustion: Progress and challenges. Renewable and Sustainable Energy Reviews. 2023;**177**:113213

[48] An X, Wang Q, Li R, Ma X, Tian Y. Comparative techno-economic analysis of waste biomass gasification for hydrogen production. International Journal of Hydrogen Energy. 2021;**46**(13)

[49] Soh J, Hailemariam S, Yoo K, Lee S, Park C, Kim J. Energy and GHG emissions of a biomass gasification system producing hydrogen for fuel cell vehicles. International Journal of Hydrogen Energy. 2018;**43**(18)

[50] Calzavara M, Finco S, Persona A, Zennaro I. A cost-based tool for the comparison of different e-grocery supply chain strategies. International Journal of Production Economics. 2023;**262**:108899

[51] Zabihian S, Sowlati T, Sokhansanj S, Bi XT, Stumborg M. Modelling and simulation of biomass gasification in fluidized bed reactors: A global kinetic scheme. Fuel. 2013;**103**

[52] Raza MA, Aqdas N, Ahmad N, Ali MI, Raza SA. Hydrogen production from biomass gasification: A review. Renewable and Sustainable Energy Reviews. 2021;**149**

[53] Piprani A et al. Supply chain integration and supply chain performance: The mediating role of supply chain resilience. International Journal of Supply Chain Management. 2020;**9**(3):58-71

[54] Lohmer J. Analysis of resilience strategies and ripple effect in blockchain-coordinated supply chains: An agent-based simulation study. International Journal of Production Economics. 2020;**228**:107882

[55] Novais L, Maqueira J, Ortiz A, Bruque S. Strategic simulation models as a new methodological approach: An application to information technologies integration, lean/just in time and lead time. Central European Journal of Operations Research. 2021;**29**:1185-1218

[56] Boubker O. The effects of information technologies on automotive supply chain and firm performance. Scientific Journal of Logistics. 2022;**18**(1):6

[57] Tukamuhabwa B et al. Supply chain resilience: Definition, review and theoretical foundations for further study. International Journal of Production Research. 2015;**53**:5592-5623

[58] Siagian H et al. Supply chain integration enables resilience, flexibility, and innovation to improve business performance in the COVID-19 era. Sustainability. 2021;**13**(9):4669

[59] Saghaei M, Ghaderi H, Soleimani H. Design and optimization of biomass electricity supply chain with uncertainty in material quality, availability and market demand. Energy. 2020;**197**:117165

[60] Liu F, Tan Y, Huang X, Sarkar S, Zhang X. Supply chain product innovation considering jointly held

options under uncertain demand. Computers & Industrial Engineering. 2023;**179**:109224

[61] Pei H, Liu Y, Li H. Robust pricing for a Dual-Channel green supply chain under fuzzy demand ambiguity. IEEE Transactions on Fuzzy Systems. 2023;**31**(1):53-66

[62] Maio F, Rem P, Balde K, Polder M. Measuring resource efficiency and circular economy: A market value approach. Resources, Conservation & Recycling. 2017;**122**:163-171

[63] Yurtay Y, Yurtay N, Demirci H, Zaimoglu E, Goksu L. Improvement and implementation of sustainable key performance indicators in supply chain management: The case of a furniture firm. IEEE Access. 2023;**11**:41913-41927

[64] Demartini M, Pinna C, Aliakbarian B, Tonelli F, Terzi S. Soft drink supply chain sustainability: A case based approach to identify and explain best practices and key performance indicators. Sustainability. 2018;**10**(10):3540

[65] Varadejsatitwong P, Banomyong R, Julagasigorn P. A proposed performance measurement system for enabling supply-chain strategies. Sustainability. 2022;**14**(19):11797

[66] Marinagi C, Reklitis P, Trivellas P, Sakas D. The impact of industry 4.0 technologies on key performance indicators for a resilient supply chain 4.0. Sustainability. 2023;**15**(6):5185

[67] Soni J, Bhattacharjee K. Multi objective dynamic economic emission dispatch integration with renewable energy sources and plug in electrical vehicle using equilibrium optimizer. Environment, Development and Sustainability. 2023

[68] Marinescu B, Bellmunt O, Dorfler F, Schute H, Sigrist L. Dynamic virtual power plant: A new concept for grid integration of renewable energy sources. IEEE Access. 2022;**10**:104980-104995

[69] Gopinathan S, Rao V, Sundaramoorthy K. Family of non-isolated quadratic high gain DC–DC converters based on extended capacitor-diode network for renewable energy source integration. IEEE Journal of Emerging and Selected Topics in Power Electronics. 2022;**10**(5):6218-6230

[70] Talaat M, Alkholy M, Alblawi A, Said T. Artificial intelligence applications for microgrids integration and management of hybrid renewable energy sources. Artificial Intelligence Review. 2023

[71] Tseng M, Ha H, Tran T, Bui T, Chen C, Lin C. Building a data-driven circular supply chain hierarchical structure: Resource recovery implementation drives circular business strategy. Business strategy and the. Environment. 2022;**31**(5):2082-2106

[72] Mahmood T, Rehman U, Naeem M. Prioritization of strategies of digital transformation of supply chain employing bipolar complex fuzzy linguistic aggregation operators. IEEE Access. 2023;**11**:3402-3415

[73] Zoric N, Maric R, Maric T, Vukmirovic G. The importance of digitalization for the sustainability of the food supply chain. Sustainability. 2023;**15**(4):3462

[74] Safarzynska K, Domenico L, Raberto M. The circular economy mitigates the material rebound due to investments in renewable energy. Journal of Cleaner Production. 2023;**402**:136753

[75] Ioannidis F, Kosmidou K, Papanastasiou D. Public awareness of

renewable energy sources and circular economy in Greece. Renewable Energy. 2023;**206**:1086-1096

[76] Romero-Castro N, Lopez-Cabarcos A, Miramontes-Vina V, Riberio-Soriano D. Sustainable energy transition and circular economy: The heterogeneity of potential investors in rural community renewable energy projects. Environment, Development and Sustainability. 2023

[77] Sahajwalla V, Hossain R. Rethinking circular economy for electronics, energy storage, and solar photovoltaics with long product life cycles. MRS Bulletin. 2023;**48**:375-385

Chapter 6

Carbon-Negative Hydrogen Production via Steam-Induced Vapolysis

Jean-Louis Kindler and Hardik Y. Desai

Abstract

The chapter focuses on the use of steam as the gasification agent in the thermochemical conversion of biomass to produce hydrogen-rich syngas and eventually 99.999% hydrogen after gas separations. Clean Energy Enterprises (CEE) is developing the Ways2H technology, to convert organic wastes into a syngas, from which hydrogen fuel and end-of-process solid co-products such as carbonates can be extracted while capturing carbon dioxide (CO_2) emissions. Waste-to-hydrogen technologies do not generate a usable energy product directly, but indirectly, by generating a hydrogen product that can be used as clean, renewable fuel in traditional power generating plants, in hydrogen-fueled transportation equipment, and for other non-fuel purposes. The Ways2H plant design generates hydrogen in a completely closed-loop thermal gasification and separation process, with minimal air emissions as compared with waste incineration technologies as well as other known gasification processes.

Keywords: vapolysis, steam gasification, hydrogen, CCUS, water-gas shift reactor, reformer, WHRB

1. Introduction

1.1 Historical background

The earliest known investigation into gasification was carried out by Thomas Shirley, who in 1659 experimented with the "carbureted hydrogen" (now called methane).

The pyrolysis of biomass to produce charcoal was perhaps the first large-scale application of the gasification-related process. When wood, owing to its overuse, became scares toward the beginning of the eighteenth century, coke was produced from coal through pyrolysis, but the use of by-product gas from pyrolysis received little attention. Early developments were inspired primarily by the need of town gas for street lighting. The silent features of town gas from coal were demonstrated to British Royal Society in 1773, but the scientists of the time saw no use for it. In 1798, William Murdoch used coal-gas to light the main building of the Soho Foundry, and in

1802, he presented a public display of gas lighting, astonishing the local population. Friedrich Winzer of Germany patented coal-gas lighting in 1804 [1].

1.2 Biomass and its products

Biomass is formed from living species such as plants and animals—that is, anything that is now alive or was a short time ago. It is formed as soon as a seed sprouts or an organism is born. Unlike fossil fuel, biomass does not take millions of years to develop. Plants use sunlight through photosynthesis to metabolize atmospheric carbon dioxide and grow. Animals grow by taking in food from biomass. Fossil fuels do not reproduce, whereas biomass does, and, for that reason, is considered renewable. This is one of its major attractions as a source of energy or chemicals.

Every year, a vast amount of biomass grows through photosynthesis by absorbing CO_2 from the atmosphere. When it burns, it releases carbon dioxide that the plants had absorbed from the atmosphere only recently (a few years to a few hours). Thus, any burning of biomass does not add to the Earth's carbon dioxide inventory. For this reason, biomass is considered a "carbonneutral" fuel.

Of the vast amount of biomass, only 5% (13.5 billion metric tons) can be potentially mobilized to produce energy. This quantity is still large enough to provide about 26% of the world's energy consumption, which is equivalent to 6 billion tons of oil.

Biomass covers a wide spectrum—from tiny grass to massive trees, from small insects to large animal wastes, and the products derived from these. The principal types of harvested biomass are cellulosic (noncereal) and starch and sugar (cereal).

All parts of a harvested crop like corn plant are biomass, but its fruit (corn) is a starch, while the rest of it is ligno-cellulose. The crop (corn) can produce ethanol through fermentation, but the ligno-cellulosic part of the corn plant requires a more involved process through gasification or hydrolysis [1].

1.3 Biomass gasification

Biomass gasification involves pyrolysis and partial oxidation in a well-controlled oxidizing environment. It leads to products, such as H_2, CO, CO_2, H_2O, and hydrocarbon species. The heat required for biomass drying, heating, and pyrolysis is provided by the partial oxidation of biomass. Gasification is deemed as the most promising technology for producing renewable and carbonfree energy, as it provides tremendous flexibility with regard to feedstock and the fuels produced. In general, the gasification process converts low-value biomass to a gaseous mixture containing syngas (mixture of H_2 and CO) and varying amounts of CH_4, and CO_2. It can also produce hydrocarbons, particularly in the lower temperature range. The oxidizing agents can be pure O_2, air, steam, CO_2, or their mixtures. The syngas composition can be varied by using air and steam as the gasification agent. Moreover, the presence of CO_2 can be used to increase H_2 and CO contents, as it transforms char, tar, and CH_4 into H_2 and/or CO in the presence of a catalyst such as Ni/Al. Numerous studies have been reported in the recent years, dealing with the type of reactors used for gasification, thermo-chemical processes involved, and various gaseous and liquid fuels produced during gasification. The syngas can be used to generate heat and power, for example, in an IGCC facility, produce H_2, and synthesize other chemicals and liquid fuels such as F-T fuels. There are various routes for the utilization of syngas, including the production of F-T and other transportation fuels. The global reactions associated with syngas formation from biomass (CHn) include the following:

$$2CHn + O_2 \Rightarrow nH_2 + 2CO \tag{1}$$

$$CO + H_2O \Rightarrow H_2 + CO_2 \tag{2}$$

$$CH_4 + H_2O \Rightarrow 3H_2 + CO \tag{3}$$

Reaction (1) corresponds to syngas formation in the presence of O_2, while reaction (2) is the well-known water-gas-shift-reaction and reaction (3) is associated with the steam reforming of methane. Reactions (2) and (3) are used to control the H_2/CO ratio. The production of F-T fuels from syngas involves a series of reactions in the presence of a catalyst. The global reactions for this process can be written as follows:

$$nCO + (2n + 1)H_2 \Rightarrow CnH_2n + 2 + nH_2O \text{ (Paraffins)} \tag{4}$$

$$nCO + 2nH_2 \Rightarrow CnH_2n + nH_2O \text{ (Olefins)} \tag{5}$$

The first step during F-T formation is the conversion of syngas into $-CH_2-$ alkyl radicals and H_2O. The $-CH_2-$ alkyl radicals then combine in a catalyst reaction to produce synthetic paraffin and olefin hydrocarbon (HC) fuels of various chain lengths. The amount and type of fuels formed are determined by parameters such as temperature, pressure, H_2/CO ratio, and the type of catalyst. In general, F-T fuels can be produced from a variety of solid, liquid, and gaseous sources, and further processed to yield clean transportation fuels with desired specifications.

Regardless of feedstock or process, F-T fuels have a number of desirable properties. For example, F-T diesel fuels can be produced with a high cetane number, with ultra-low sulfur and aromatic content, with the consequence of improved engine performance, significantly lower particulate mass (PM) emissions and favorable NOx/PM trade-off. However, these fuels generally have poor lubricity and lower volumetric energy density. These shortcomings can be alleviated by blending these fuels with petro-fuels. Thus, the biomass gasification can be used to produce syngas and subsequently clean drop-in transportation fuels. The effects of F-T fuel properties on engine performance and emissions have been reported by a number of investigations [2].

1.4 Gasification medium and types of gasification agents

A gasifying medium assists to break down the heavier solid hydrocarbons in order to convert them into low-molecular-weight gases such as CO and H_2. There are a few well-known gasifying agents used in the process of biomass gasification, namely air, steam, oxygen, and carbon dioxide.

The influence of using different agents on energy efficiency has been studied extensively in the literature [3, 4]. For instance, Sharma et al. [3] conducted an experimental study on biomass gasification process in a downdraft reactor followed by saturated steam when the steady-state conditions achieved. They indicated that adding steam to air in the reduction zone increased the hydrogen flow rate. Ismail and El-Salam [4] studied air gasification of biomass in an updraft reactor and indicated that equivalence ratio, which is a function of both air and biomass, influenced the composition of synthesis gas. They concluded that the equivalence ratio has an indirect relation with the concentration of CO and H_2 while interacting directly with the CO_2 content of the gas [3, 4].

1.5 Types of gasifiers

There are several types of biorefinery reactors that impact the thermochemical conversion, for example, updraft, downdraft, cross-draft, fluidized bed, and entrained flow gasifier which fall into the three main categories: (1) moving (fixed) bed, and (2) fluidized bed and (3) entrained flow gasifier. Moving bed and fluidized bed gasifiers are discussed here.

1.6 Moving bed gasifiers

Updraft gasifiers are the most straightforward configuration in moving bed reactors. In a conventional updraft gasifier, fuel is fed from the top, while the syngas leaves from the top. The gasifying agent (air, oxygen, steam, or their mixture) is preheated and fed into the gasifier through a grid at its bottom.

In a downdraft gasifier, biomass is fed from the top, while the entered gasification agent meets with the pyrolysis product, releasing heat. After that, both gas and solid (char and ash) products move down in the downdraft gasifier. Here, a part of the pyrolysis gas may burn above the gasification zone; this phenomenon, called as flaming pyrolysis, supplies the thermal energy required for the endothermic reactions through the combustion of pyrolysis gas. Drawbacks of a downdraft gasifier are namely grate blocking, channeling, and bridging which hinder scaling-up the gasification process. The moisture content of feedstock in a downdraft gasifier should not exceed 30% to avoid inferior products and low efficiency. A downdraft gasifier is beneficial for producing low-tar syngas and has a simple operation configuration. The syngas of a downdraft gasifier has less tar as well as lower LHV compared to an updraft [5].

1.6.1 Fluidized bed gasifier

Unlike other types of biorefinery reactors mentioned earlier, fluidized-bed gasifier contains non-fuel granular solids (bed solids) that act as a heat carrier and a mixer. The two types of fluidized bed are circulating fluidized bed and bubbling fluidized bed, differing in fluidization techniques. In a bubbling fluidized bed, the fuel fed from either the top or the side is mixed quickly throughout the fluidized bed. The gasifying medium serves as the fluidizing gas and is sent through the bottom of the reactor, leading to a quicker mixture of fuel particles with the bed materials and thus much faster heating of the fuel. More rapid drying and pyrolysis processes occur in the bubbling fluidized bed gasifiers compared to the circulating type. A deficiency of this system is that partially gasified particles can exit the process as a result of mixing with gasified solids. The combustion reaction occurs in the fluidized phase resulting in a lower efficiency [5].

1.7 Typical gasifier design

Gasification is a relatively old technology that has been in use throughout the industrialized world for approximately 200 years. During the nineteenth century, "town gas" was produced from coal and peat and used as an industrial fuel and municipal gas supply for street and home lighting. The first manufactured gas plant in North America was opened in Baltimore, Maryland, in 1816. Since the 1920s, gasification has been used for the synthetic production of chemicals (methanol, ammonia, etc.). Being petroleum-poor but coal-rich, Germany used gasification coupled with the Fischer-Tropsch process during World War II to produce liquid fuels. Fischer-Tropsch production accounted for an estimated 9% of German war production of fuels and 25% of the automobile fuel supply. Similarly, in South Africa, another country with large coal reserves but little oil, SASOL started commercial production of diesel and gasoline in 1952 using coal gasification and the Fischer-Tropsch process.

Gasification can be thought of as a partial combustion process where steam reacts with the carbon in a solid carbonaceous fuel or feedstock *via* the general reaction:

$$C + H_2O - > CO + H_2 \tag{6}$$

This reaction occurs at high temperatures (800–1000°C, 1472–1832°F). It is an endothermic process, meaning that heat is consumed by the reaction and a source of heat is required to support it. In a traditional gasifier, this heat is generated by providing just enough air or pure oxygen to achieve the desired gasification temperature in order to sustain the gasification reactions. That is the partial oxidation process. This combustion burns fuel/feedstock via the following general combustion reactions:

$$C + 1/2O_2 - > CO \tag{7}$$

$$CO + 1/2O_2 - > CO_2 \tag{8}$$

$$H_2 + 1/2O_2 - > H_2O \tag{9}$$

There is an inherent trade-off between the desired gasification temperature and the amount of feedstock that is consumed to generate the heat required. Partial oxidation "consumes" feedstock and as such reduces the potential for free hydrogen production [5].

Interestingly, at high temperatures another well-known chemical reaction, named the water-gas shift (WGS) reaction, occurs:

$$CO + H_2O < - > CO_2 + H_2 \tag{10}$$

This is a reversible reaction that will eventually reach an equilibrium condition where the rate of reaction to the right equals the rate of reaction to the left. The equilibrium composition of CO, H_2O, CO_2, and H_2 is a function of the amount of carbon, oxygen, and hydrogen available and system temperature. According to the reaction stoichiometry shown in Eq. (5), if CO_2 could be selectively removed, then the reaction would shift to the right, producing more H_2. Minimizing the production of CO_2 via combustion helps to maximize the production of hydrogen, which leads back to the inherent trade-off between a desirable gasification temperature and the amount of feedstock that is "sacrificed" to maintain this temperature.

A gaseous mixture of CO, H_2O, CO_2, and H_2 is commonly referred to as synthesis gas (syngas). A variety of follow-on technologies are used to convert syngas to methane, methanol, acetic acid, gasoline, diesel, waxes, etc. Syngas can also be burned cleanly in a burner or fuel cell and purified hydrogen can be produced using a pressure swing absorption (PSA) process to selectively remove hydrogen from the syngas.

The most common gasifier types are shown in **Figure 1**. Note that these gasifier types involve a single reaction vessel where all reaction zones occur in the single vessel.

A typical downdraft gasifier is presented in **Figure 1(a)**. The reactor volume consists of a bed of solids starting with fresh feedstock fed at the top converted to

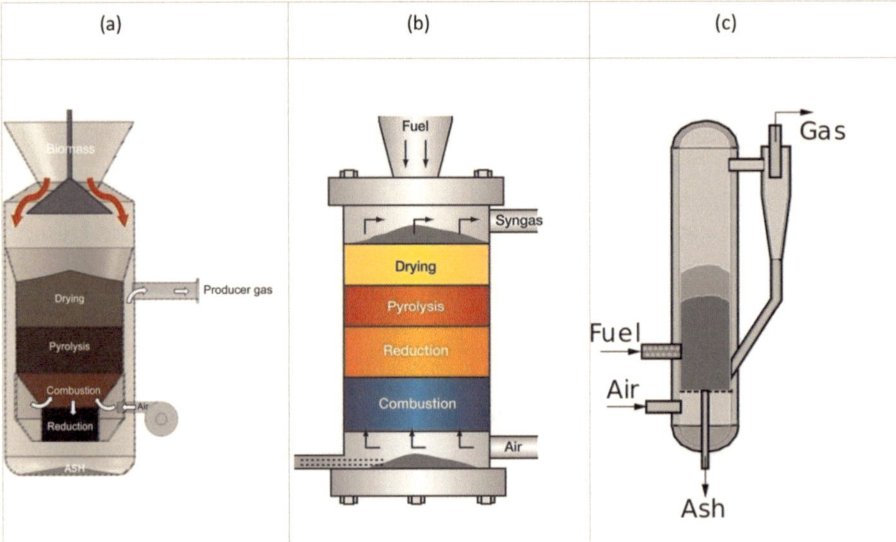

Figure 1.
Common gasifier types. (a) Downdraft gasifier. (b) Updraft gasifier. (c) Fluidized bed.

Reactor Zone	Temp. Range	Heat of Reaction
Drying	> 150 °C	Endothermic
Pyrolysis	150–700°C	Endothermic
Combustion	700–1500°C	Exothermic
Reduction	800–1000°C	Endothermic

Table 1.
Reaction zones in a downdraft gasifier.

residual char at the bottom. The feedstock moves down the reactor under the influence of gravity. **Table 1** lists the reaction zones that occur in a downdraft gasifier and the range of temperatures for each zone.

As feedstock particles are dried, pyrolyzed, partially combusted, and gasified (reduced), the particles decrease in size and the bed density increases. A small amount of air is introduced with the feedstock at the top of the gasifier, but the majority of air is introduced in the combustion zone. Heat released by combustion facilitates pyrolysis and drying by radiant heat transfer. The char leaving the combustion zone is heated to gasification temperatures. The syngas moves concurrently with the solid bed and is drawn off after the reduction zone.

Ash and char fall through the bed grate and are collected from the vessel bottom.

In **Table 1**, the heat of reaction is listed for each zone. Drying, pyrolysis, and reduction reactions are mostly endothermic (heat is absorbed), while the combustion zone is exothermic (heat is released). It is interesting to note that drying temperatures exceed the normal boiling point of water (100°C @ atmospheric pressure).

Forcing water out of the feedstock fibers is a diffusion process that requires more energy than simply the heat of vaporization of water.

An updraft gasifier, **Figure 1(b)**, is similar to a downdraft gasifier except that all air is introduced at the bottom of the gasifier and the produced syngas leaves at the top of the reactor. Notice, due to where the air is introduced, the relative locations of the combustion and reduction zones flip. At the bottom of the gasifier, the residual gasification char is burning to heat up the rising gas as well as provide radiant heat for the higher reaction zones.

A fluidized bed gasifier, **Figure 1(c)**, uses a non-combustible bed material such as sand to transfer heat and facilitate mixing in a bubbling or fully fluidized regime. Feedstock enters the reactor, coming in contact with hot sand. The air and evolving syngas are at a high enough velocity to cause the sand and feedstock to rise. As the feedstock rises up the reactor, it undergoes the various stages of gasification. A gas-solid cyclone is used to separate the produced syngas from the fluidizing medium. In a bubbling bed, velocities are maintained so that the majority of the sand will stay in the bottom half of the reactor. In a fully fluidized bed, the gas velocity is high enough to carry almost all of the sand overhead where it is recovered by the cyclone and returned to the bottom of the reactor.

Indirect heating presents the advantages of causing little or no consumption of feedstock, making most of the fuel (feedstock) available for the gasification reactions. However, the management of the sand in the reactors can become problematic as it raises issues with energy consumption to maintain the flow, as well as maintenance concerns resulting from abrasion due to contact between sand and the reactor wall.

1.8 Gasifier reaction chemistry

A general reaction network for feedstock gasification is presented in **Figure 2**.

Wet feedstock is dried by hot rising gases and/or radiant heat from the combustion zone. Usually, feedstock is dried externally to 10–20 wt% moisture before being used as a feed to a gasifier. At this level of dryness, the water that remains is bound in the fibers and interstitial spaces of the feedstock. Energy is required to increase the diffusion rate of water from the tightly wrapped fibers. This explains why drying occurs at temperatures above the normal boiling point for water.

As the temperature of the feedstock approaches 150 to 180°C, the feedstock begins to breakdown into smaller molecular compounds, including gases, tars, and char. The tars are aromatic and phenolic compounds in their vapor form. For wood and agricultural residues, the glucose and pentose rings in cellulose and hemicellulose, and the phenol groups in lignin yield these tars. The final char mostly consists of carbon and represents the fixed carbon that is analyzed to characterize it as a solid fuel (proximate analysis).

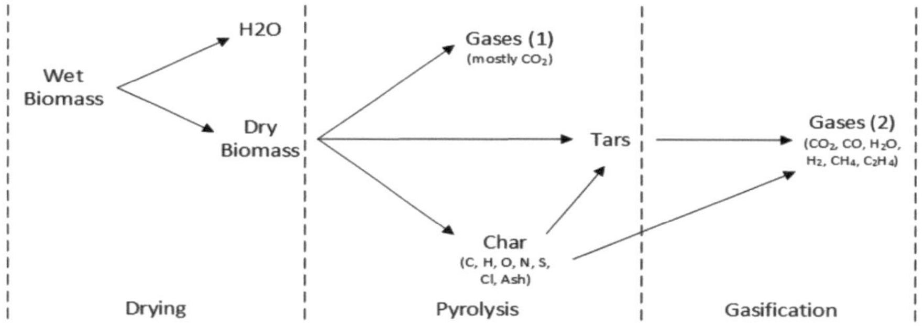

Figure 2.
Generalized reaction network for feedstock gasification.

Reactions	Reaction Stoichiometry	$\Delta H_{reaction}$ (MJ/kmol)	Reaction Name
R-1	$C(s) + H_2O(g) \rightarrow CO(g) + H_2(g)$	+131	Steam Gasification
R-2	$CO(g) + H_2O(g) \leftrightarrow CO_2(g) + H_2(g)$	−41	Water-Gas Shift
R-3	$CH_4(g) + H_2O(g) \rightarrow CO(g) + 3H_2(g)$	+206	Steam Methane Reforming
R-4	$CnHm(g) + H_2O(g) \rightarrow nCO(g) + (n + m/2) H_2(g)$	< 0	Steam Tar Reforming
R-5	$C(s) + 2H_2(g) \rightarrow CH_4(g)$	−75	Hydrogasification
R-6	$CO(g) + 3H_2(g) \rightarrow CH_4(g) + H_2O(g)$	−227	Methanation
R-7	$C(s) + CO_2(g) \rightarrow 2CO(g)$	+172	Boudouard
R-8	$CnHm(g) + nCO_2(g) \rightarrow 2nCO(g) + (m/2) H_2(g)$	> 0	Dry Tar Reforming

Table 2.
Gasification reactions.

Pyrolysis involves a complicated set of thermal cracking reactions that are difficult to characterize beyond the empirical treatment of the reaction network shown in **Figure 2**. The overall gasification (reduction) reactions, presented in **Table 2**, are more easily characterized. These reactions consist of three groups: (1) reactions involving steam, (2) reactions involving hydrogen, and (3) reactions involving carbon dioxide.

Notice, that the gasification reactions are actually a mix of endothermic ($\Delta H_{rxn} > 0$) and exothermic ($\Delta H_{rxn} < 0$) reactions. However, on balance, the reduction zone is net endothermic.

Table 3 lists the overall combustion (oxidation) reactions where $C_nH_m(g)$ represents tar species. All oxidation reactions occurring in the combustion zone are exothermic. The combustion heat is transferred by heating up the local gas and solids as well as by radiant heat transfer to the other sections of the gasifier.

1.9 Waste to hydrogen: general process

A process targeting carbonaceous matter conversion into hydrogen will require a combination of specific features that include (but not limited to):

- Little or no partial oxidation of the feedstock
- Constant, high reaction temperatures to maximize hydrogen production and avoid tar persistence.
- Controlled presence of water vapor

All waste to hydrogen projects must begin with feedstock preparation, which includes sorting, sizing, storage, and feeding. Feeding can be the most difficult as the feedstock must be isolated from the hot process and fed to the process in a controlled

Reactions	Reaction Stoichiometry	$\Delta H_{reaction}$ (MJ/kmol)	Reaction Name
C-1	$C(s) + 1/2O_2(g) \rightarrow CO(g)$	−111	Carbon Partial Oxidation
C-2	$CO(g) + 1/2O_2(g) \leftrightarrow CO_2(g)$	−283	CO Oxidation
C-3	$H_2(g) + 1/2O_2(g) \rightarrow H_2O(g)$	−242	H_2 Oxidation
C-4	$C_nH_m(g) + n/2O_2(g) \rightarrow nCO(g) + m/2H_2(g)$	< 0	Tar Oxidation

Table 3.
Combustion reactions.

flow, particularly when it contains a substantial number of plastics, which can melt and cause mechanical issues in the feeding system.

The feedstock is fed into one of several types of pyrolyzers or gasifiers: fluidized bed or fixed bed type. These are typically heated to 600–650°C either directly through combustion or indirectly via intermediate thermal media, such as sand or ceramic balls. Combustion obviously burns part of the feedstock, which includes the product. Indirect heating has the partial heat losses associated with heating an intermediate media, which then heats the feedstock and the heat losses, as well as that increases CAPEX, of significant additional equipment. The syngas exits from the pyrolyzer and the thermal media and char exit the bottom and are separated. The thermal media is then conveyed up to the preheater where is it cross exchanged with the flue gas from the process heater, which burns the char and the off-gas from the hydrogen separator. With partial oxidation solutions, the challenge lies in producing as little combustion byproducts as possible, particularly when processing complex waste mixes such as municipal solid waste.

The syngas from the pyrolyzer is then typically reformed to 1000–1100°C by the combustion of part of the syngas with the addition of oxygen. Reforming to 1100°C does crack some of the light hydrocarbons, tars, and fine char while reforming reactions occur.

The reformed gas is then treated to cool the gas via a quench tank, waste heat boiler, and scrubber. The reformed gas is then treated by an amine system or specialty scrubbers to remove the gases such as: ammonia, hydrogen cyanide, hydrogen sulfide, hydrochloric acid, carbonyl sulfide. Once those contaminants are removed, the reformed gas is processed through a Pressure Swing Absorber (PSA) to separate the carbon monoxide and carbon dioxide from the hydrogen, the final product. The hydrogen recovery from the PSA is about 72%. The remaining 28% of the hydrogen and the reject or off-gas from the PSA can then be reused either internally in the system, or for additional third-party processing, such as a gas engine.

2. Ways2H process description

2.1 General principles

As discussed above, Ways2H recognized the disadvantages of the past processes and competitive technologies in detail, as well as some innovative commercial technologies that could be applied to waste to hydrogen conversion. In applying these technologies to the design of the Ways2H process, the object was to simplify the process, eliminate certain equipment, decrease operating expense, and increase hydrogen yield.

Some solutions use heat carriers as the pathway for feedstock indirect heating. While this solution seems to bear several advantages, implementing a heat carriers circuit within the process is also source of complexity; therefore, cost increases. Eliminating the use of thermal media, such as the heat carriers above as well as fluidized bed removes the need for the following:

1. A thermal media preheater that cross-exchanges the media with flue gas from the process heater,
2. A process heater or combustor itself which is a source of emission of NOx and wasted energy,
3. A mechanical conveyor which transports the media,
4. A char separator and conveyor necessary to separate the char from the media after pyrolysis,
5. Expensive and complex valves systems,
6. In summary, a great deal of capital and operating cost.

Another process type recently evaluated was plasma gasification. In this process, the heat required for feedstock gasification is provided via generation of a plasma, typically reaching temperatures in the range of 4000°C. In addition to significantly high CapEx, these solutions require a high maintenance, particularly with the reactors' internal walls, subject to relatively high temperatures compared with conventional Vapolysis or pyrolysis, as well as the electrodes that generate the arc for plasma generation. Made of graphite, these electrodes are essentially expensive consumables.

Then of course, there are several processes on the market using more conventional solutions like partial oxidation of raw feedstock as the heat source. There are well-known cases where these processes, which are very effective for predictable, stable feedstock streams such as wood or coal, have been plagued by mechanical incidents, mostly caused by tar occurrence, due to too low temperatures in the gasification/pyrolysis reactors.

The challenge then is to provide enough heat to ensure full gasification without combusting the hydrogen-generating components of the feedstock with a mechanically simple configuration. In order to accomplish full gasification without the complexity and cost of an intermediate media, an alternate source of energy is required for the thermal cracking reactions—vapolysis—to occur. The solution is based on the choice of a high-temperature heat source for the reformer, and then to utilize the hot syngas from that reformer as the heat source for vapolysis. A device producing a reforming temperature approximately at 1400°C was identified.

However, the challenge was to utilize the heat from the reformed gas, avoid the use of solid thermal media, and yet to keep the reformed gas separate from the gas in the vapolysis process. Two technologies were evaluated: an indirectly heated rotary kiln or an indirectly heated screw conveyor. Rotary kilns are typically used to heat solids to the point where a chemical reaction or physical change takes place, in this case: drying, devolatilization, and vapolysis, which makes the kiln's temperature control critical. A rotary kiln is comprised of a rotating cylinder (called the drum) and jacket, which surrounds the drum (sometimes called the furnace), sized specifically to meet the temperature and retention time requirements of the material to be processed. The kiln is set at a slight angle, in order to allow gravity to assist in moving the feedstock through the rotating cylinder. Mechanical rotation movement also assists in breaking the eventual feedstock clusters that may arise. The problem arises in the mechanical lip seals and bridging between the furnace and rotating drum. These seals are mechanically difficult to build and maintain. In addition, the leakage of these seals or their failure led to highly flammable gases leaking to the atmosphere [6].

2.2 Main features of the process

Screw Vapolyser is a modified version of an indirect heated screw conveyor. It has several advantages:

1. Initially, feedstock is shredded and pre-processes by several mechanical operations to suit the quality of feedstock with Screw Vapolyser processing. The feedstock is shredded up to inches of size to have smooth gasification by increasing the reaction rate.

2. The feedstock is thoroughly mixed by the mechanical action of the screw itself, promoting the vapolysis process.

3. Steam is added in Screw Vapolyser as gasification agent.

4. The rotating screw also increases the heat transfer to the feedstock as well increases the reaction efficiencies between steam and feedstock, which helps to have efficient gasification.

5. The feedstock is gasified in the lower portion of the screw and the syngas is separated from the solids in an upper zone.

6. A suction blower pulls the syngas from the Vapolyser into gas mixer and reformer for those gaseous reactions that generate more hydrogen. The unconverted portion of feedstock, which is mainly inorganic in nature, comes out as bottom ash at the end of the screw.

7. Reformer is refractory lined reactor with 1400°C operating temperature. Such a high temperature is achieved using part of the hydrogen produced as fuel through oxy-fuel burner (OFB).

8. The 1400°C reformed gas from the reformer provides the gasification energy for Vapolysis at very little CapEx or OpEx.

9. The 1400°C reformed syngas flows through an enclosure, which surrounds the screw conveyors tube. There are no seals except on screw shaft, which uses a conventional rotary seal.

10. The temperature differential between the reformed gas and the feedstock/ Vapogas is much greater than any intermediate media solution.

11. The reformed gas, after having released its certain portion of heat to the Screw Vapolyser, is cooled down, cleaned, dried and undergoes a final conversion phase, Water Gas Shift Reaction, where carbon monoxide is removed to produce more hydrogen. Prior to entering the Water Gas Shift Reactor (WGSR), gas must be heated up to 250 to 300°C, as per particular operating temperature requirement, for better conversion and reaction rates in WGSR.

12. Separation phases yield:-

    a. Hydrogen product, of which a fraction is diverted to feed the burner in the HT reformer

    b. Carbon dioxide, which is sent to an integrated mineralization component. The output from this component is mineralized carbon, permanently sequestered and available for example as a construction material or for concrete/cement manufacturing processes (**Figure 3**) [6].

Feeding the Vapolyser, the organic feedstock is first preheated with excess process steam to remove excess moisture in the feedstock if any. The feedstock then enters the Vapolyser through a rotary gas lock, which not only meters the feedstock into Vapolyser but seals the gases being generated in the Vapolyser, preventing them from flowing back into the feedstock line. As the feedstock passes through the Vapolyser, it is exposed to a controlled temperature, which gradually increases the feedstock from its initial—ambient—temperature to final temperature of 850°C, which is much higher than gasifiers using intermediate media.

The vapolysis temperature of 850°C is achieved by transferring the reformed gas heat at 1400°C, directly, through the jacket enclosure of the Vapolyser. This would cause the vapolysis process to occur while raising the feedstock temperature to 850°C. At this point, with exception of a minimal remaining solid carbon fraction (char) and mineral fraction (ash), the feedstock is totally converted to syngas, essentially its elemental gases, such as hydrogen, carbon monoxide, carbon dioxide, methane, and

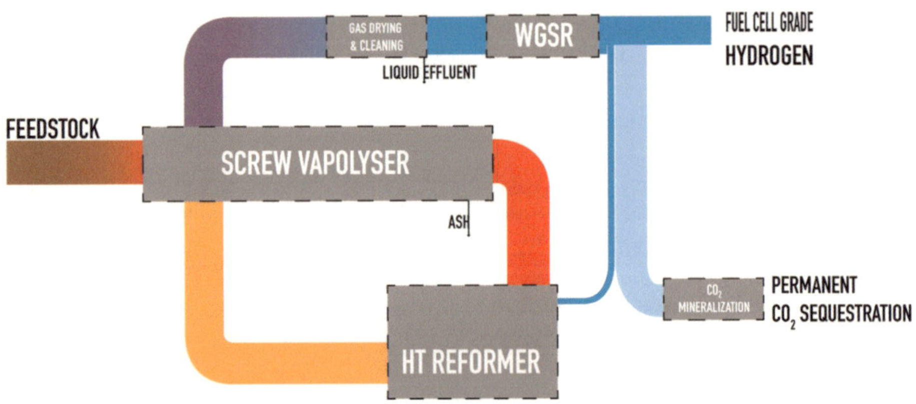

Figure 3.
Overall Ways2H process schematic.

small amounts of contaminants such as hydrochloric acid, hydrogen cyanide, ammonia, COS, and hydrogen sulfide. The syngas then flows to the reformer with any remaining entrained tars after passing through the "Gas mixer/Static Mixer" where more steam is added stoichiometrically required by the reforming process. The only waste product is a minimal amount of ash that may result from solid char decomposition. This ash will be removed from the bottom of the Screw Vapolyser.

The syngas and char enter the reformer where they are exposed to an Oxy-Fuel Burner (OFB), which exposes the Vapogas and any leftover tar from the Vapolyser to temperatures of 1400°C. Additional steam provision is there in the reformer if required. In the OFB, oxygen, in stoichiometric proportion, is mixed with fuel, which is 99.999% pure hydrogen from the Hydrogen Storage/Hydrogen Separator product stream.

Combustion of this fuel yields a flame of over 2000°C and temperatures in excess of 1400°C in the reformer. At this temperature, essentially all of the methane and tars are cracked to hydrogen and carbon monoxide, increasing the hydrogen content over a conventional reformer by about 10 to 15%. In addition, contaminant gases such as ammonia and hydrogen cyanide are converted to nitrogen and water, eliminating NOx emissions.

The OFB hot gas (combustion products) accelerates to sonic velocity through a specifically designed nozzle, forming a turbulent jet. This turbulent jet has an enormous entrainment appetite that enhances mixing between the Vapogas containing tar/methane and the reactive hot steam. The combination of high temperature and rapid mixing makes the jet extremely reactive. Therefore, the OFB process can provide better syngas (defined as CO and H_2) yields than if conventional oxygen injection solutions were used in a typical reformer.

In addition to the existing mixing effect directly caused by the OFB nozzle, Ways2H has developed and integrated a proprietary static mixing component, which through an innovative use of aerodynamics principles further increases the gas flow reactivity and provides an improved mixing of the injected steam with the Vapogas flow before entering the reformer.

The combined reformed gas and OFB combustion gases exit the reformer and enter the various sections of the vapolysis jackets through temperature control valves. Each valve controls the flow of 1400°C reformed gas to each jacket section according to the desired temperature for each stage of the vapolysis. There is a control on the flow entering the vapolysis heating jackets, which allows extra hot reformed gas to bypass the heating jacket and is directed to the Waste Heat Recovery Boiler (WHRB).

After exiting the vapolysis jackets, the reformed gas then enters heat recovery, cleaning, and separation phase. The reformed gas from the jacket of the Vapolyser is at approximately 1000°C and enters waste heat recovery boiler, which generates process steam for preheating the feed and injection into the Vapolyser, Gas Mixer/Static Mixer, and reformer. The reformed gas then enters a quench tank, which cools the reformed gas. The reformed gas then enters a dual scrubber further cools the gas and completes the removal of contaminants. After scrubbing down acidic gases, a fine particulate filter (<0.5 micron) helps removing all particles still present and gases will be free of acids and particles at this stage. A compressor increases the reformed gas pressure and electric heater increases the temperature required to feed the Water-Gas Shift Reactor (WGSR).

Water-Gas Shift Reactor (WGSR) is used to convert carbon monoxide (CO) into carbon dioxide (CO_2) and hydrogen (H_2) with the help of added steam. This step is very useful to enhance the hydrogen production.

The gas then enters a carbon dioxide separator and removes almost all carbon dioxide, which is then fed to the Carbon Capture & Mineralization (CCM) system. This system mineralizes the carbon dioxide with Ca+, Mg+, CaO, $Ca(OH)_2$, MgO, Mg $(OH)_2$ from various natural as well as a by-product sources from other industrial processes to form calcium, sodium, or magnesium carbonate. The advantages of the CCM technology over traditional Carbon Capture Storage (CCS) technology are the conversion of CO_2 to a stable, storable, and usable carbonate compound. Eventually, all the carbon entering the process in the feedstock as well from the burners/gas engine will find its way to CO_2 and ultimately mineralize as carbonate, making the whole process a full carbon sink.

At this stage, the gas is free of carbon dioxide (CO_2) and has majorly hydrogen (H_2), carbon monoxide (CO), nitrogen (N_2), and traces of Hydrogen Sulfides (H_2S)/ other sulfur compounds. This gas mixture is all set to be processed through Hydrogen Separator, which requires further sulfur removal and concentration below 50 PPB. To achieve the limit on sulfur presence, sulfur removal system is used to remove sulfurs prior to be treated by Hydrogen Separator.

The OFB nozzle uses 99.99% Hydrogen (H_2) to generate 1400°C reformed gas, which in turn provides the energy for feedstock gasification in the Screw Vapolyser and the auxiliary burner (if required) use the off-gases or waste gases from the process to provide heat to Screw Vapolyser if hot-reformed gas heat supply is not sufficient to bring up and maintain the temperature of Screw Vapolyser up to 850°C.

There is an exhaust from the Hydrogen separator system, which is still having mainly carbon monoxide (CO), Hydrogen (H_2), and Nitrogen (N_2). This exhaust stream can be used either by Aux-Burner for Screw Vapolyser if required, and if not, this stream can be used by Gas Engine/SOFC to generate electricity for captive use and as discussed above, the exhaust of electricity generation system, having carbon monoxide (CO_2), again diverted to CCM system for mineralization [6].

2.3 The Ways2H process equipment

In order to achieve the above process, Ways2H utilizes a series of commercially proven processes or components that are operated within their design limits.

The first stage of Vapolysis takes place in a Vapolyser, which is a heated shaftless screw conveyor.

The typical Vapolyser with a feedstock inlet, a Vapogas outlet to the Reformer, a bottom outlet for the inorganic ash fraction of the feedstock, a heating jacket surrounding the conveyor, and an external motor.

The hot gas from the Vapolyser then enters the Reformer where it is heated to 1400°C by a direct flame. The 1400°C gas from the Reformer then passes through the Vapolyser jacket, indirectly heating the feedstock to 850°C, achieving the vapolysis of the feedstock to its elemental gases of mainly hydrogen, carbon dioxide, carbon monoxide, and methane. The 850°C Vapogas temperature is controlled by the flow of gas from the reformer to the jacket. The Ways2H 850°C Vapogas is much hotter than the 600°C pyrolysis gas typical in the industry, achieving more complete gasification of tars and char industry typically calls this mixture of gases, syngas (**Figure 4**).

The conveyor screw's, shown here, main function is to move the organic feedstock along the conveyor tube to provide uniform mixing, cluster breaking, and better heat transfer from the conveyor tube to feedstock. In these applications, the materials of construction are designed to withstand temperatures of over 1000°C with alloys such

Figure 4.
Internal view of Screw Vapolyser.

as: Nitronic 60 or Inconel 625 type. The conveyors like this are used by many similar applications, including in the pyrolysis and calcination industries.

Before further processing, the high-temperature Vapogas is sent through a Particulate Filter in order to remove the particles above 1 micron size. The particulate filter is a metallic mesh-type filter provided by several suppliers. The second phase of Vapolysis takes place in a Gas Reactor where the Vapogas and additional steam are injected via static mixers. The Gas Reactor is sized to allow sufficient residence time for certain gas reactions to occur.

The third and final stage of Vapolysis occurs in the Reformer.

Reformer is a process used in almost all petroleum industries. Gas reformers in one or other way process the raw gas to achieve highest concentration of targeted product. Steam reformers are the current mainstream pathway for hydrogen production, with methane (natural gas) as feedstock. As such, it is also a well-known, widely used technology.

In the third and final phase in the Ways2H Vapolysis process, Vapogas from the Vapolyser passes into a High Temperature Reformer to cause certain gas reactions to occur. In order to achieve 1400°C exhaust gas that enters the Vapolyser jacket, the Ways2H Reformer uses a commercial size oxy-fuel burner, which burns a small fraction of the ultimately produced hydrogen with oxygen, achieving much higher temperatures and the breakdown of methane to hydrogen.

The high-temperature flame heats the Vapogas to over 1400°C—from that step it is called Hot Reformed Gas. The Hot Reformed Gas then passes through the jacket of the Vapolyser, achieving an optimal efficiency of thermal energy utilization.

The process has now completed the Vapolysis stages and commences the Reformed Gas treatment and separation processes. The first stage is the cooling of the gas and the generation of steam for the various gas reactions.

After Hot Reformed Gas passes through the Vapolyser jacket, its temperature is about 1000°C. It then enters a conventional waste heat recovery boiler. The boiler generates steam for injection into the Vapolyser, Reformer, and Water Gas Shift Reactor. The steam is actually a source of added hydrogen for the reactions that occur in these vessels. In addition, the gas is cooled for subsequent processes.

The waste heat recovery boiler concept is already a well-known and proven technology. There are several WHRB Designer/Manufacturers/Suppliers available globally.

The cooled gas then passes into a standard Dual Scrubber, which removes low levels of contaminants such as: HCl, NH_3, HCN, and COS, which were introduced in the feedstock as trace components. Those trace elements must be removed to allow the gas to be effectively separated in downstream systems and not to contaminate or poison media or membranes in downstream systems. Gas scrubbers are commonly available systems, which will be selected based on our requirement as per the pollutants' concentration in our process gas.

The entire system starting from Screw Vapolyser to Gas Scrubbers is operated under slight vacuum. An ID fan/Suction fan at the end of Gas scrubber is responsible for maintaining the gas flow and pressure from throughout up to ID fan/suction fan outlet.

Induced Draft Fan/Suction fans are impeller-based motor driven fans, and the speed is controlled by various differential pressure points at various processes. These type ID fans, well-designed according to the Ways2H process gas and operational requirement, are commonly available.

The treated gas is then pressurized with a typical dual acting reciprocating compressor and heated in heat exchanger prior to entering the Water Gas Shift Reaction (WGSR). The WGSR requires a certain pressure and temperature for efficient operation.

The WGSR is a widely used industrial reaction for the manufacture of ammonia, hydrocarbons, methanol, and hydrogen. In the manufacture of hydrogen, the carbon monoxide is mixed with steam in the presence of certain catalyst to convert the carbon monoxide to carbon dioxide and additional hydrogen. After removing the contaminants that could poison the catalyst, the Reformed Gas in the Ways2H process gas, which still has a significant concentration of carbon monoxide, is processed to produce significantly more hydrogen with minimal energy consumption.

After the Reformed Gas passes through the WGSR, it consists mostly of carbon dioxide, hydrogen, nitrogen from the feedstock and trace sulfur contaminants. The next step in purifying the gas to hydrogen is to separate the carbon dioxide.

Pressure Swing Adsorption (PSA) or Vacuum Pressure Swing Adsorption (VPSA) or any other commercially proven and available membrane and absorption technologies are two well-known variants of a technique used to separate some gas species from a mixture of gases under pressure/vacuum/normal pressure according to the species' molecular characteristics and affinity for an adsorbent material. It operates at near ambient temperature and significantly differs from cryogenic distillation, another process commonly used to separate gases.

For the removal or separation of carbon dioxide, Ways2H uses proven CO_2 separators technologies for the highest efficiency. The separated carbon dioxide then flows to the CCUS system, which is discussed later.

The retentive gas from the CO_2 separator is primarily hydrogen with nitrogen from the feedstock and small amounts of carbon dioxide and carbon monoxide. Prior to the final separation of fuel cell grade hydrogen, sulfur compounds that were not removed in the scrubber must be removed to prevent poisoning the palladium in hydrogen separation membranes.

Before the hydrogen can be separated from the treated gas, it must be pressurized and heated, similar to the conditioning process used before the WGSR, using a second dual acting reciprocating compressor and heat exchanger.

The final step in the production of fuel cell grade hydrogen is to remove the nitrogen due to the feedstock and small amounts of carbon dioxide and carbon monoxide left from the WGSR and CO_2 separators. To achieve the fuel cell grade hydrogen, the treated feed gases are passed through a commercial palladium membrane system to separate the fuel cell quality hydrogen from the waste gas. The waste gas along with the nitrogen is then burned to provide an additional thermal energy source for the Vapolyser or the WGSR heat exchanger or hydrogen separator heat exchanger. A small fraction of the hydrogen product is burned in the reformer's oxy-fuel burner.

The final major product of the Ways2H process is the Carbon Capture Use & Storage (CCUS) of the separated carbon dioxide from the CO_2 Separator. The carbon dioxide is then sent to mineralization process for carbon sequestration into carbonate/solid form.

Bottom line, Ways2H achieves the conversion one dry ton per day of organic waste to a net output of approximately 150 kg per day of fuel-cell grade hydrogen using conventional, industrially proven, and commercially available equipment operating in a unique temperature and sequential profile [6].

3. Economics

The technical aspects of the process have been outlined above, showing the particularly promising performances in terms of hydrogen yield, when calculated on the basis of feedstock input.

Technical performance does not provide much benefit if the solution cannot be economically deployed and operated. To estimate the economic feasibility of the solution, we chose to focus on two main indicators, one being the overall economic performance, expressed as project Internal Rate of Return (IRR). The other performance indicator is the hydrogen production cost. The following estimation is based on a system being built and operated on a generic site in the continental United States, with basic CapEx and OpEx numbers provided as follows.

3.1 Capital expenditure (CapEx)

Table 4 below outlines the main cost elements of a fully delivered, commissioned turnkey plant. The retained perimeter (battery limits) starts with processed feedstock injection, and ends with products (hydrogen and byproducts) output, conditioning thereof not included.

3.2 Operational expenditure (OpEx)

As most industrial processes, the plant requires several utility connections to operate. **Table 5** below summarizes all considered costs for system operations. It is important to note that in an alternative configuration, the plant can be equipped with an internal stationary fuel cell to produce its own electricity and as such be able to operate off-grid. It is also worth to note that a significant part of OpEx resides in the CO_2 mineralization function, particularly in precursor procurement.

3.3 Overall economic performance

Overall economic performance is a complex calculation, based on several concomitant factors, whose combination provides a specific business model and profitability

CapEx (Capacity in tpd)		**24**
Average estimated H_2 production per day (kg)		3600
Equipment cost estimation		
Purchased equipment		
Concrete		
Controls		
Electrical		
Fire Protection		
Heat Trace & Insulation		
Mechanical		
Rentals		
Site work		
Structural - Equipment		
Structural—Pipe Supports		
System Install & Commissioning		
Miscellaneous		
Other Direct costs		
Direct Costs		**19,524,890**
Peripheral Equipment		
Preprocessing	No	—
Fuel Cell	No	—
Compressor (bar)	No	—
Other additional Equipment	No	—
Contingencies	5%	976,245
Complete system	**24 tpd**	**20,501,135**
Initial Project Assessment		90,000
Basic Engineering		424,645
Detailed Engineering		2,406,321
Project supervision & Management	2%	410,023
Insurance	2%	410,023
Indirect costs & Engineering		**3,741,012**
Total cost before grants		**24,242,146**

Table 4.
Delivered system CapEx.

calculation. Each specific project should be assessed on the basis of local conditions such as utility prices, feedstock supply (negative) cost, logistics costs, as well as products and byproduct sale prices.

OpEx	Cost basis		Quantity	$/year
Maintenance				
System maintenance, per year	3%			475,602
Maintenance total				475,602
Utilities				
Energy consumption (kW)	$0.15		1355 kW	1,576,149
Reformer Oxygen cost	$0.15 /kg			189,212
Process Water	$0.41 /kg		6393 m3/y	2621
Material processing cost (preprocessing)	$0.00 /kg		9690 t/y	—
Ca source incl. Logistics	$90.00 /ton		31,913 t/y	2,872,151
Total Utilities				4,640,132
Disposal				
Ash and other solids disposal	$30.00 /ton		4772 t/y	143,168
Total Disposal cost				143,168
Labor	Monthly		Count	
System Manager	$	3000	1	18,000.0
Shift in charge	$	1500	4	72,000.0
Maintenance technician	$	2000	1	24,000.0
Feedstock handling labor	$	1500	1	18,000.0
Total weighted—yearly			132,000	132,000
Payroll Taxes & Benefits		15%	19,800	19,800
Total payroll yearly expense			151,800	151,800
Total OpEx per year				**5,410,703**

*Notes to **Table** 5 above:*
Oxygen cost is an internally calculated average of various hydrogen sources delivered cost.
Cost of processed water is based on US California average cost in 2019.

Table 5.
System OpEx.

3.3.1 Hydrogen production cost

Hydrogen is the main output and purpose of these solutions. Although this hydrogen will likely be compressed for use or transportation, the present calculation is based on hydrogen output before compression, but delivered at PEMFC-compliant quality.

A primary characteristic of waste-to-hydrogen systems—as opposed, for example, to electrolytic hydrogen—lies in the multiplicity of revenue streams. While being more complex to manage, these also help mitigating the operation risk, as revenue variations on one parameter will have a lower impact on overall operations. Other revenue streams that are being accounted for are, by order of importance:

- Mineralized CO_2 (in this example, calcium carbonate)
- Carbon credits
- Waste processing fees (tipping fees)

Tipping fees reference																			$ 50
Carbon Credits reference																			$ 60
		Tipping fees																	
	$ 0.19	$ (20)	$ (10)	$ -	$ 10	$ 20	$ 30	$ 40	$ 50	$ 60	$ 70	$ 80	$ 90	$ 100	$ 110	$ 120	$ 130	$ 140	$ 150
	$ -	1.98	1.88	1.77	1.67	1.57	1.47	1.37	1.27	1.16	1.06	0.96	0.86	0.76	0.66	0.55	0.45	0.35	0.25
	$ 10	1.80	1.70	1.59	1.49	1.39	1.29	1.19	1.09	0.99	0.88	0.78	0.68	0.58	0.48	0.38	0.27	0.17	0.07
C	**$ 20**	1.62	1.52	1.42	1.31	1.21	1.11	1.01	0.91	0.81	0.71	0.60	0.50	0.40	0.30	0.20	0.10	−0.01	−0.11
a	**$ 30**	1.44	1.34	1.24	1.14	1.03	0.93	0.83	0.73	0.63	0.53	0.43	0.32	0.22	0.12	0.02	−0.08	−0.18	−0.29
r	**$ 40**	1.26	1.16	1.06	0.96	0.86	0.75	0.65	0.55	0.45	0.35	0.25	0.14	0.04	−0.06	−0.16	−0.26	−0.36	−0.46
b	**$ 50**	1.08	0.98	0.88	0.78	0.68	0.58	0.47	0.37	0.27	0.17	0.07	−0.03	−0.14	−0.24	−0.34	−0.44	−0.54	−0.64
o	**$ 60**	0.90	0.80	0.70	0.60	0.50	0.40	0.30	0.19	0.09	−0.01	−0.11	−0.21	−0.31	−0.42	−0.52	−0.62	−0.72	−0.82
n	**$ 70**	0.73	0.62	0.52	0.42	0.32	0.22	0.12	0.02	−0.09	−0.19	−0.29	−0.39	−0.49	−0.59	−0.70	−0.80	−0.90	−1.00
	$ 80	0.55	0.45	0.34	0.24	0.14	0.04	−0.06	−0.16	−0.26	−0.37	−0.47	−0.57	−0.67	−0.77	−0.87	−0.98	−1.08	−1.18
C	**$ 90**	0.37	0.27	0.17	0.06	−0.04	−0.14	−0.24	−0.34	−0.44	−0.55	−0.65	−0.75	−0.85	−0.95	−1.05	−1.15	−1.26	−1.36
r	**$ 100**	0.19	0.09	−0.01	−0.11	−0.22	−0.32	−0.42	−0.52	−0.62	−0.72	−0.83	−0.93	−1.03	−1.13	−1.23	−1.33	−1.43	−1.54
e	**$ 110**	0.01	−0.09	−0.19	−0.29	−0.39	−0.50	−0.60	−0.70	−0.80	−0.90	−1.00	−1.11	−1.21	−1.31	−1.41	−1.51	−1.61	−1.71
d	**$ 120**	−0.17	−0.27	−0.37	−0.47	−0.57	−0.67	−0.78	−0.88	−0.98	−1.08	−1.18	−1.28	−1.39	−1.49	−1.59	−1.69	−1.79	−1.89
i	**$ 130**	−0.35	−0.45	−0.55	−0.65	−0.75	−0.85	−0.95	−1.06	−1.16	−1.26	−1.36	−1.46	−1.56	−1.67	−1.77	−1.87	−1.97	−2.07
t	**$ 140**	−0.52	−0.63	−0.73	−0.83	−0.93	−1.03	−1.13	−1.24	−1.34	−1.44	−1.54	−1.64	−1.74	−1.84	−1.95	−2.05	−2.15	−2.25
s	**$ 150**	−0.70	−0.80	−0.91	−1.01	−1.11	−1.21	−1.31	−1.41	−1.52	−1.62	−1.72	−1.82	−1.92	−2.02	−2.12	−2.23	−2.33	−2.43
	$ 160	−0.88	−0.98	−1.08	−1.19	−1.29	−1.39	−1.49	−1.59	−1.69	−1.80	−1.90	−2.00	−2.10	−2.20	−2.30	−2.40	−2.51	−2.61
	$ 170	−1.06	−1.16	−1.26	−1.36	−1.47	−1.57	−1.67	−1.77	−1.87	−1.97	−2.08	−2.18	−2.28	−2.38	−2.48	−2.58	−2.69	−2.79
	$ 180	−1.24	−1.34	−1.44	−1.54	−1.65	−1.75	−1.85	−1.95	−2.05	−2.15	−2.25	−2.36	−2.46	−2.56	−2.66	−2.76	−2.86	−2.97

Table 6.
Hydrogen production cost sensitivity analysis.

DOI: http://dx.doi.org/10.5772/intechopen.1003604

			Year 1	Year 2	Year 3	Year 4	Year 5	...	Year 25
Annual Tons of Waste Processed (dry tons)	(1)	7752							
Revenue:									
Total Annual Revenue			**13,510,277**	**13,519,967**	**13,529,851**	**13,539,933**	**13,550,216**	...	**13,805,065**
Operating Costs:									
Total Operating Costs			**5,824,504**	**5,879,962**	**5,935,975**	**5,992,547**	**6,049,686**	...	**7,320,400**
EBITDA		57%	**7,685,774**	**7,640,005**	**7,593,876**	**7,547,385**	**7,500,530**	...	**6,484,665**
Depreciation and Amortization	20%	10 years	$1,939,372	$1,939,372	$1,939,372	$1,939,372	$1,939,372	...	$0
Earnings before Interest and Taxes			5,746,402	5,700,634	5,654,505	5,608,013	5,561,158	...	6,484,665
Interest Expense		8.00%	1,315,826	1,219,977	1,116,173	1,003,752	882,001	...	0
Earnings before Taxes			4,430,575	4,480,656	4,538,332	4,604,261	4,679,157	...	6,484,665
Estimated Taxes	(10)	25.00%	1,107,644	1,120,164	1,134,583	1,151,065	1,169,789	...	1,621,166
Tax Credit	per kg H_2 produced	$ -	—	—	—	—	—	...	—
Earnings after Taxes			3,322,932	3,360,492	3,403,749	3,453,196	3,509,368	...	4,863,499
Debt Payment (principal only)	(11)		1,154,816	1,250,665	1,354,470	1,466,890	1,588,642	...	(0)
Net Cash Flow			2,168,115	2,109,827	2,049,279	1,986,305	1,920,726	...	4,863,499
Cumulative Cash Flow			(5,104,529)	(2,994,702)	(945,423)	1,040,883	2,961,609	...	90,037,671
Project Internal Rate of Return	20 years	31%							

Table 7.
25-year financial statement summary.

As an example, **Table 6** shows the sensitivity analysis of hydrogen production cost, based on tipping fees and carbon credit variations and calculated over a 25 years operation period.

The average unweighted tipping fees in continental USA are calculated at US$54.17. For convenience, we have rounded this number down at $50. Similarly, there are over 170 carbon credit calculation types and bases available. World Bank [7] states a range between $40 and $80 in 2021. We chose a median price of $60. This combination results in a hydrogen production cost of $0.19 per kg, compression not included [6].

3.3.2 Internal rate of return

Costs have been described in the OpEx table above. In addition, CapEx amortization, project financing costs as well as other costs such as taxes also need to be taken into account when assessing the final economic model and profitability of a project. **Table 7** below provides a summarized overview of a standard project financial statement.

As shown in the bottom line of the table, operations result in an IRR over 20 years of 31% [6].

4. Conclusion

In summary, the Steam-Induced Vapolysis process as designed by Ways2H does the following:

1. Although being a fully integrated Waste to Hydrogen solution, with CCUS included, the process eliminates a significant number of expensive and complex equipment. The direct impact is multiple:

2. Significant reduction in system CapEx

3. Increase in heat efficiency with a significant reduction in OpEx.

4. Better reliability using modified commercially available technologies.

5. Increases hydrogen yield from the use of high temperature of vapolysis and reforming along with Water-Gas Shift Reactor (WGSR) addition, significantly improving operation economics.

Excellent control is provided over the vapolysis temperature gradient and overall level with the specific combination of multi-sectional regions.

A concentrated source of carbon dioxide is provided for mineralization to carbonates for sales and significant carbon credits.

Is characterized by the use of proven technologies in similar applications. Their combined use in the Ways2H process is unique.

When deployed on-site for commercial operations, a generic system, operating under average parameters will produce hydrogen at a cost of 0.19 per kg, with an overall Internal Rate of Return of 31% (**Figures 5** and **6**).

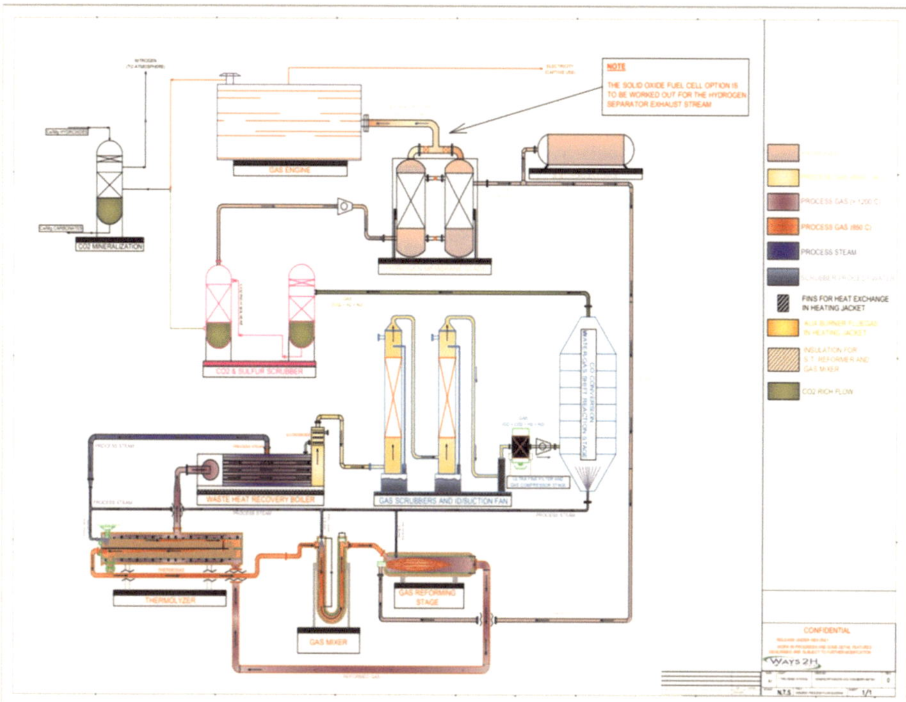

Figure 5.
Process flow diagram.

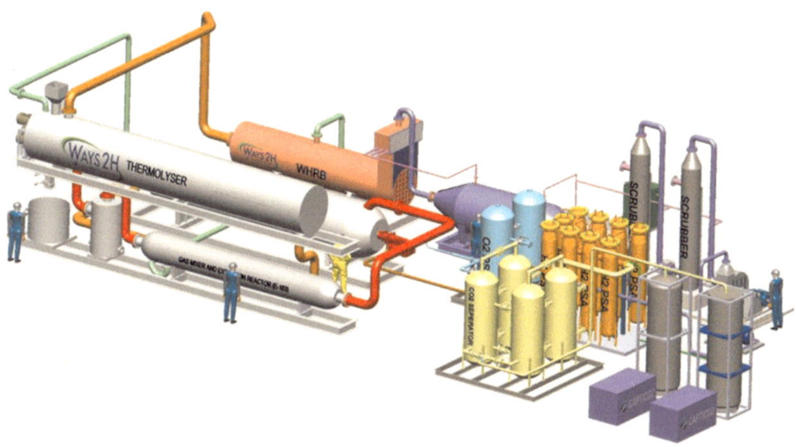

Figure 6.
Plant layout.

Author details

Jean-Louis Kindler* and Hardik Y. Desai*
Ways2h, Inc. Long Beach, CA, USA

*Address all correspondence to: jlkindler@gmail.com and hardikydesai@gmail.com

References

[1] Basu P. Biomass Gasification and Pyrolysis – Practical Design. Academic Press; 2010. pp. 1-5

[2] Bundschuh J. Technologies for Converting Biomass to Useful Energy. Vol. 4. Toowoomba, Australia: University of Southern Queensland (USQ); 2013. p. 10

[3] Sharma S, Sheth PN. Air–steam biomass gasification: Experiments, modeling and simulation. Energy Conversion and Management. 2016;**110**: 307-318

[4] Ismail TM, El-Salam MA. Parametric studies on biomass gasification process on updraft gasifier high temperature air gasification. Applied Thermal Engineering. 2017;**112**:1460-1473

[5] Mazaheri N. Modeling the Gasification Process of Wood-Char Biomass. Ste-Anne-de-Bellevue, QC, Canada: Department of Bioresource Engineering, McGill University; 2018. pp. 7-10

[6] Ways2H [Internet]. Available from: http://www.ways2h.com [Accessed: January 11, 2024]

[7] World Bank. State and Trends of Carbon Pricing 2021. Washington, DC: World Bank; 2021. Available from: http://hdl.handle.net/10986/35620